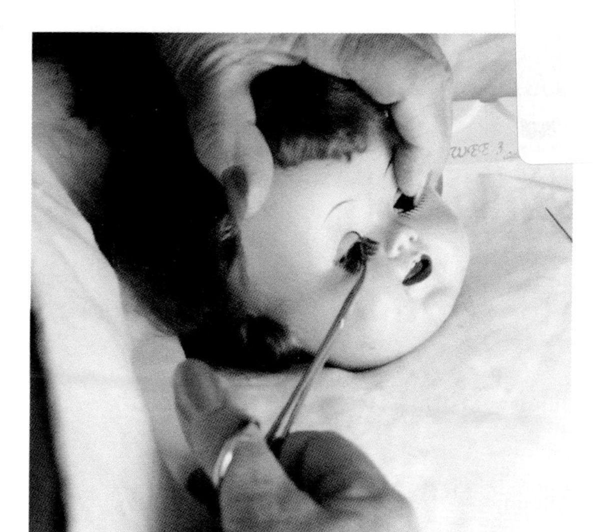

All About Doll Repair & Care

A Guide to Restoring Well-Loved Dolls

by Carol Lindberg

Second printing 2006.

Printed in the United States.

ISBN: 1-58597-383-1

Library of Congress Control Number: 2006923011

A Note from the Author:

Because of the many composition doll restoration and repair methods practiced by doll hospitals and restoration and repair experts, as well as the numerous cleaning and restoring products available today for dolls, the author urges the reader to proceed with care in using any of these methods and products on any doll restoration or repair project. It cannot be emphasized too strongly that any product needs to be first tried on the doll in an inconspicuous place to make sure it will do no further damage to the doll. According to professional museum curators as well as restoration and repair experts, the most important tenet to remember is that no restoration or repair work should ever be done to an object which is nonreversible, and any repair should be done with materials as close to the original as possible. This same tenet also applies to doll restoration and repair.

The information in this book is presented in good faith and every effort has been made to ensure that all information is accurate. All recommendations are made without guarantee on the part of the author or publisher. The author and publisher disclaim any liability in connection with the use of this information and cannot be held responsible for any injuries, losses, or other damages that may result from the use of information or procedures in this book.

Additional copies of this book and other doll care products are available from:

www.allaboutyourdolls.com

4500 College Boulevard
Overland Park, Kansas 66211
1-888-888-7696
www.leatherspublishing.com

Dedication

This book is dedicated to my grandchildren:
Brian James Brown
Larry Dale Brown, Jr.
Michael Dale Brown
Sarah Michelle Smith
Shelley Marie Smith
Rick Edward Winfrey
and
Great Granddaughter Ashley Marie Smith

The family of Carol Lindberg: Husband Tom, Daughters Janet and Ginnie, and son Larry.

Carol Lindberg — four generations.

Acknowledgments

I would like to thank the following individuals for their encouragement and support, without which this book would never have been completed:

My granddaughter, Sarah Smith, for her illustrations.

My youngest daughter, Ginnie Cary, and my son-in-law, Larry Smith, for the hours of work editing and typing the manuscript.

My husband, Tom Lindberg.

My oldest daughter, Janet Smith.

My son, Larry Brown.

All of my grandchildren.

Barbara Frohlich of Dollspart Supply in New Jersey, for loaning her *Miss Revlon* doll to me.

Table of Contents

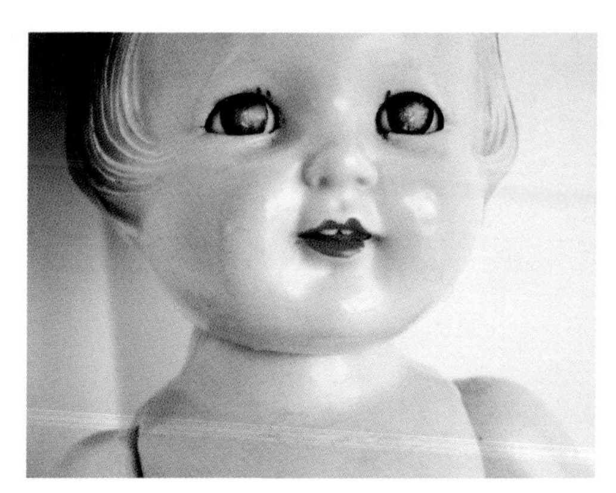

Doll Eyes 85

Doll Clothing 93

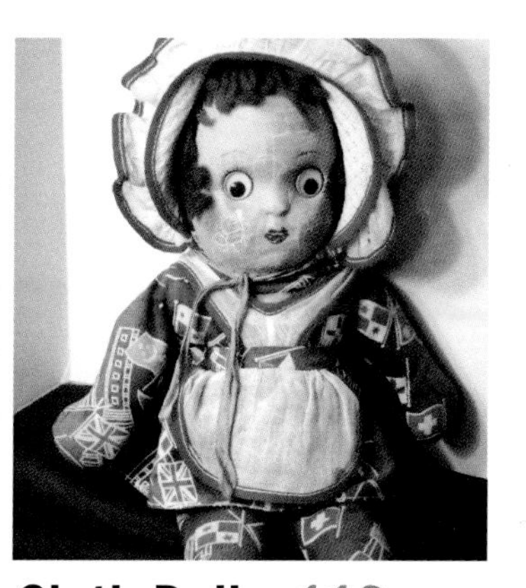

Cloth Dolls 113

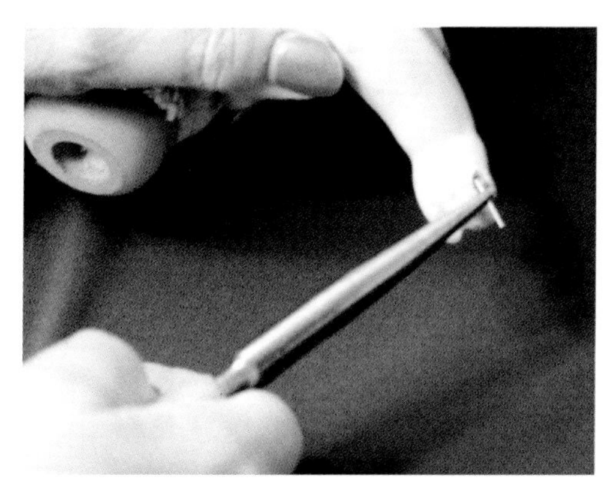

Composition Dolls 117

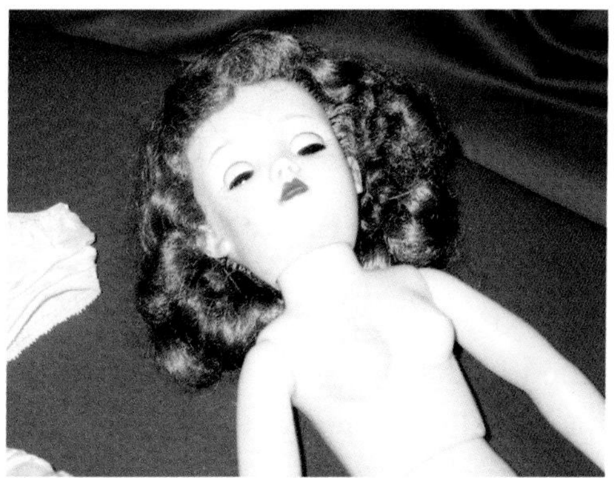

Tips for Certain Kinds of Dolls 125

Foreword

Doll restoration started out as a hobby for me years ago, but has since developed into a passion. Realizing the value of all the information accumulated over the years on restoring dolls and methods which have been successful, the decision was made to present them in an easy-to-use format. Here you will find instructions for following procedures developed through years of experience as well as tricks-of-the-trade which have been learned along the way.

Whether you are a novice at doll restoration or an accomplished restorer, it is my hope that you will find this book enjoyable as well as instructive. A novice will discover a new phase in the world of doll collecting while the more experienced restorer may learn some new methods to be used in restoration.

The methods described in this book are what I have found work for me – after trial and error. In sharing them with you, it is hoped you will be spared the trial and error part of doll restoration. As you proceed, you may come up with other methods which will help to simplify your task. Experience is one of the best teachers!

Remember to enjoy your work! What could be more pleasurable for someone who loves dolls than to be able to help them return to their original glory! And, in the process, you will be providing pleasure and enjoyment to the dolls' owners, not to mention the satisfaction you will feel yourself for a job well done!

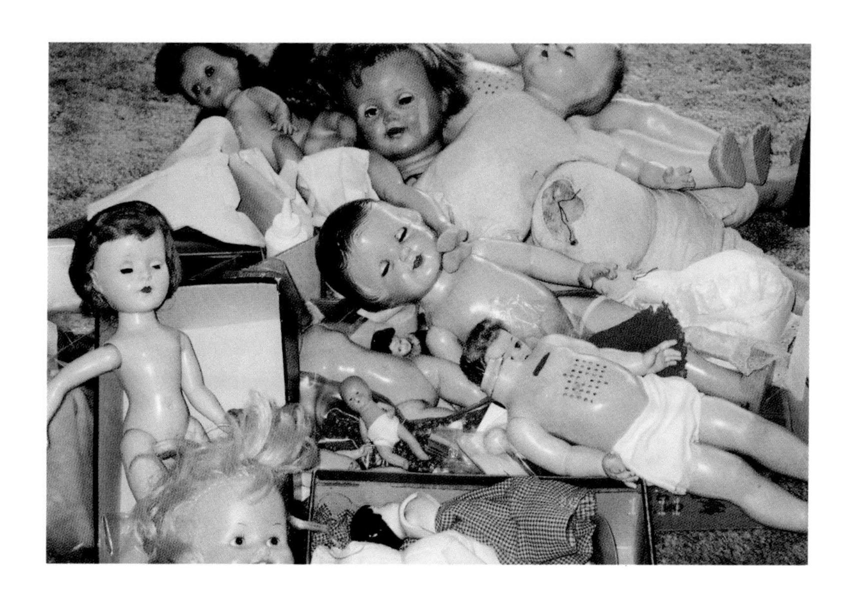

Introduction

So, you want to restore dolls! Whether you have purchased this book to learn to restore dolls of your own and those you may collect in the future, **or** if you are inspired to start your own business in this field, you have made a good choice! The information contained in this book will be of valuable assistance to the amateur restorer as well as the more professional restorer.

Before getting into the actual restoration of a doll, there are some guidelines that should be followed.

1. **Practice!** I cannot stress enough that you should purchase every doll you see at a garage sale, thrift shop, white elephant sale, flea market or elsewhere. Buy every one that you can afford no matter what the condition of the doll is at the time. (**SEE PHOTOS ON THIS PAGE AND PAGE 6.**) Believe me, you will make use of all of them in time and it will be a big help to have them on hand. Take a good look at them and select one that appeals to you. Take it to your work area and start the restoration process. This is how you gain "experience."

2. **Practice!** Scrutinize the doll very carefully and study how it is put together. You will find this will help you develop the skills to restring or put together different types of dolls. Also, this will be very valuable to you when you start restoring dolls for others, as you will already know how long the process will take, what supplies are necessary to complete the process and what price to set for the work.

3. Finally, **practice** — on dolls that are of little value monetarily or emotionally. It is better to learn techniques on such dolls rather than those in which you have a personal stake or one in which your customer does. A rule of thumb is do not do something to your doll or to someone else's doll that cannot be undone as you do not want to make a bad situation worse. It is better to try your techniques on a doll that could not get worse even if you are unsuccessful at your restoration.

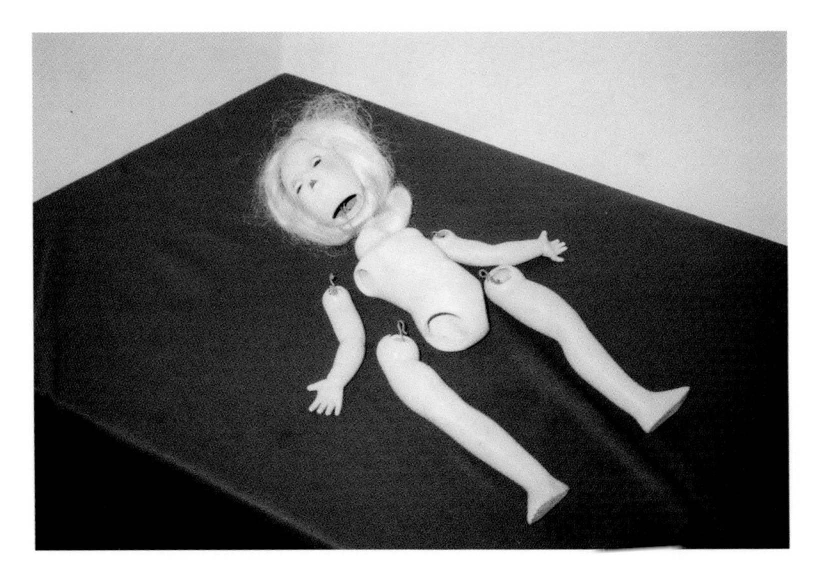

Take the doll completely apart.

The Basics

The first thing to stress is that the"tips" contained in this book are what work for me. You are welcome to try them, but at your own risk. There is a possibility that "restoration" can actually do more harm than help. Each situation is unique as no two dolls age exactly the same way. I suggest, in order to be safe, that you try each technique in an inconspicuous place on each doll or piece of clothing with which you are working.

Normally, all old (and some new) dolls need to be restrung and most of them will need to be cleaned as well. If so, cut the elastic or whatever was holding the doll together. Take the arms, legs, head and torso apart. You will find they are much easier to work with one piece at a time. (SEE PHOTO ABOVE.)

The first step in doll care is cleanliness. There really is no value in "dirt." Many things are taken into consideration in valuing a doll and dirt is not one of them! If anything, a dirty doll decreases in value. However, you must be careful that cleaning a doll to remove the dirt does not do more harm than good, as a damaged doll is valued at even less than a dirty one!

Inspect each piece to determine what type of work or repairs are needed and how best to proceed. The limbs may have metal hooks that are broken or rusted. If so, replace them. If the doll still has the original hooks, then you already know the size of the replacements. If the original hooks are in good condition, they may be used again. If it is apparent that they have been replaced at some time in the past with the incorrect size (or with paper clips, rubber bands or whatever else was used), this is the time to remedy the situation.

A limb may need to be repaired in order to restring the doll. Of course, that will have to be done prior to restringing. You will learn as you go along and thus gain experience. Most problems can be repaired by somehow attaching the hooks. Perhaps that means

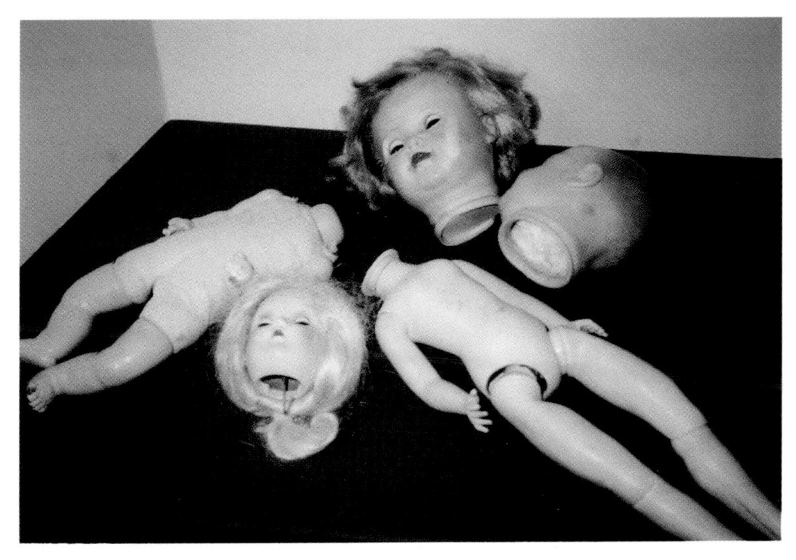

Various doll heads.

drilling a new hole or using putty to build up the joint so a new hole can be made. Many times this repair will be entirely inside the doll once she is restrung. The putty is not useful on vinyl dolls, as it will not stick.

Auto body putty is the easiest kind of putty I have found to use. It can be purchased at almost any auto parts supply in a quart can much like paint. You will also need a tube of hardener. Read the instructions on the can and ask the salesman for his tips. It is very easy to use, and through experience you will learn just how much to mix up at a time. One rounded tablespoon of the putty and a drop or two of the hardener mixed together fixes a lot of problems. It needs to be used immediately, so have everything ready when you mix it. Smooth it out as much as possible in the area you are restoring. However, after it hardens, it is still quite easy to sand. I find wood putty is much harder to work with.

Inspect the head and neck area. Doll heads are attached in different ways; however, most will have something that enables them to be attached to the rest of the body. In bisque heads, there is usually a neck button that protrudes from the neck so the elastic can be attached to it and then to the rest of the body. Composition and hard plastic dolls usually have a rod inside the head just behind the ears or a "bale" of wire attached to the head that is visible and you can easily to put the hook on it to attach the elastic. (SEE PHOTO ABOVE.) China heads or bisque heads with shoulder plates are a little different (see the chapter on "Doll Bodies and Their Repair"). Some vinyl dolls have flanges of the same vinyl that hold the limbs and heads to the torsos and are not strung with elastic at all. A hair dryer is a wonderful tool to use to warm the joints so they are pliable enough to remove or reattach without splitting the plastic.

Remember, before restringing all the parts, the doll needs to be cleaned.

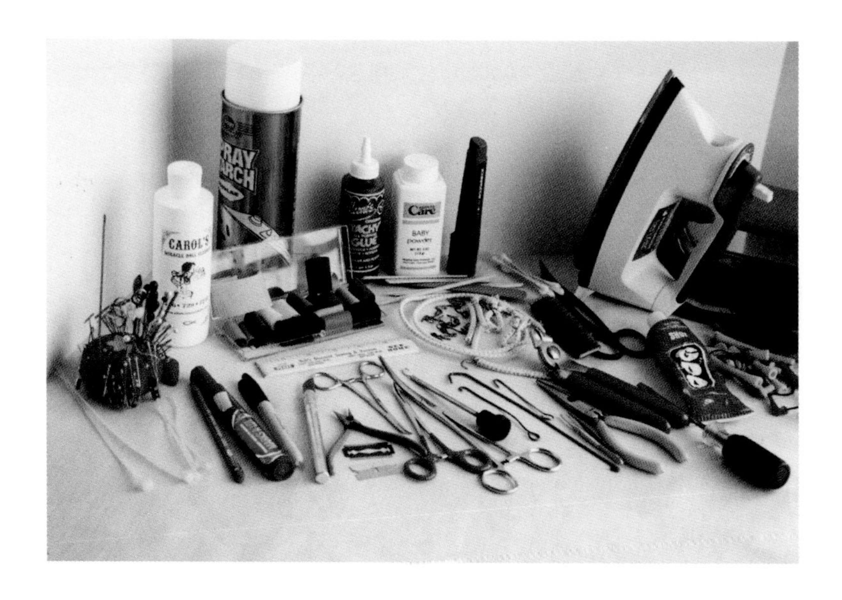

Tools and Supplies for Your Work Area

This listing of tools and supplies in- cludes some you will need as you begin and others you will need as you advance in your doll restoration work. Many of these items can be found around the house, some you will need to purchase specifically for use in restoration and, for others, it is recommended that you acquire them as you need them.

- Acrylic fingernail material
- Adhesive tape
- Auto body putty
- Baby powder
- Band-Aids®
- Bleach: Clorox® is recommended
- Bobby pins
- Bottlebrush
- Cleaning tool – used to clean ceramics before firing
- Clothespins
- Combs
- Cotton swabs
- Craft sticks
- Crochet hooks
- Dryer sheets
- Elastic (for stringing dolls)
- Emery board fingernail file
- Fan (for comfort)
- Fabric Finish: Magic Sizing is recommended
- Fabric softener in liquid form: Downey is recommended
- Felt-tipped permanent pens: black and brown ink
- Glue: Aleene's Designer Glue™, nail glue, Shoe Goo, super glue
- Good light
- Hairbrushes
- Hair dryer
- Hair spray: MINK™ brand is recommended
- Hand steamer
- Hemostats in several sizes (scissors-like clamps)
- Hog rings
- Iron
- Iron-on tape: several colors but lots of beige or tan
- Margarine tub: two-pound size
- Measuring tape - flexible cloth type
- Moto tool
- Needles: sewing needles and a very long 8-9in (20-23cm) needle
- Oil: Sewing machine oil or 3-in-1 oil
- Old kid gloves: preferably white or off-white, for body repair
- Old towels and washcloths
- Palmolive® dishwashing soap
- Paint

The author's work area.

- Paper clips: various sizes including the large size
- Permanent curlers and papers
- Pink setting wax
- Pins: safety pins, straight pins, long corsage or hatpins
- Plaster of paris
- Pliers: needle nose
- Puff iron
- Razor blades: both double edge and single edge
- Ruler
- "S" hooks
- Sand paper
- Scissors: good ones and an old pair for cutting elastic and wire
- Screw driver (a large one to pry with)
- Sealer (for newly painted area)
- Sleeve board
- Spray bottle
- Spray starch
- Steam iron
- Toothpicks: wooden
- Tweezers
- Wig spray
- Wire — #24 (or close to it)
- X-acto knife
- Yardstick

In addition to these tools and supplies, you will need **boxes and boxes of doll parts**. You will find these are your saving grace in many situations. I have even "cut" a vinyl arm off to the length I needed to put on a cloth body! This vinyl arm was on a completely vinyl doll and was just the right size and shape to "fix" the doll with a missing arm. It was pinned in place in the cloth body and then taken to a shoe repairman who sewed it in place! Another doll was made "whole" again! (Instructions to accomplish all of this are given in the chapter on "Doll Bodies and Their Repair.")

You will also need a place to work. My "workbench" has always been my ironing board where I can sit while working and also have a surface on which to work. When my "workroom" was moved from our home to the shop and back again, my ironing board, sewing machine and serger were always in a "U" shape with a swivel secretarial chair in the center. Shelves containing supplies surrounded the area and were within easy reach. My sewing machine cabinet was opened at all times, allowing for my tools to be close at hand. (SEE PHOTO ABOVE.)

Getting Started In Business

Once you have accomplished any restoration work, the calls will start coming in. Your neighbor learns that you have restored your childhood doll and brings her doll over to see if you will do the same for her! Your cousin calls and says she heard you were now restoring dolls for other people and she has three she wants done for Christmas presents. Your little girl (or granddaughter) has a friend with a doll whose leg is partially ripped off of the cloth body and tape is not holding it anymore.

The requests will continue to come in and before you know it, you realize you could be in business. At this point you need to determine if you want to do doll restoration as a business, and, if so, you need to get better organized. If your decision is to set up a business, there are some steps you definitely need to follow.

The first thing you will need is a Sales and Receipt Book. (You can purchase these at any office supply store. They will even order them for you imprinted with your business name and address when you are ready for that.) You will need the one that comes in *triplicate*.

In the presence of the owner/customer, complete the date, name, address and telephone number of the owner/customer. Even though the owner is a friend/relative/acquaintance, it is still important to have this information in writing as you may need it later when you least expect it! If the doll has a name, record that also. Then when you speak of the doll to her, now and later when the doll is ready to go home, you have "personalized" your transaction

Still in the presence of the owner/customer – and this is **very important** – write down a complete description of the doll (type of doll, hair color, eye color, complete list of clothing and accessories). (SEE PHOTOS ON PAGES **13 & 14.**) Look the doll over carefully, taking off all the clothes while you are doing so. This is for your own protection. If there is damage under her clothing, both you and the owner will see it at this time, and you should comment on anything good or bad at that very moment.

Ascertain what needs to be done – restringing, new hairdo, eyes brightened, doll cleaned entirely. At this time, dip a cotton swab in Carol's Miracle Doll Cleaner and rub it on the doll in an inconspicuous area to be certain it does not damage the doll. If this is successful, you can try it on another area where there is more dirt, such as the face or arms. Your customer will see how clean her beloved doll can be and will immediately want you to clean the entire doll.

Examine each piece of clothing as well. Talk about

Miss Revlon doll and clothing.

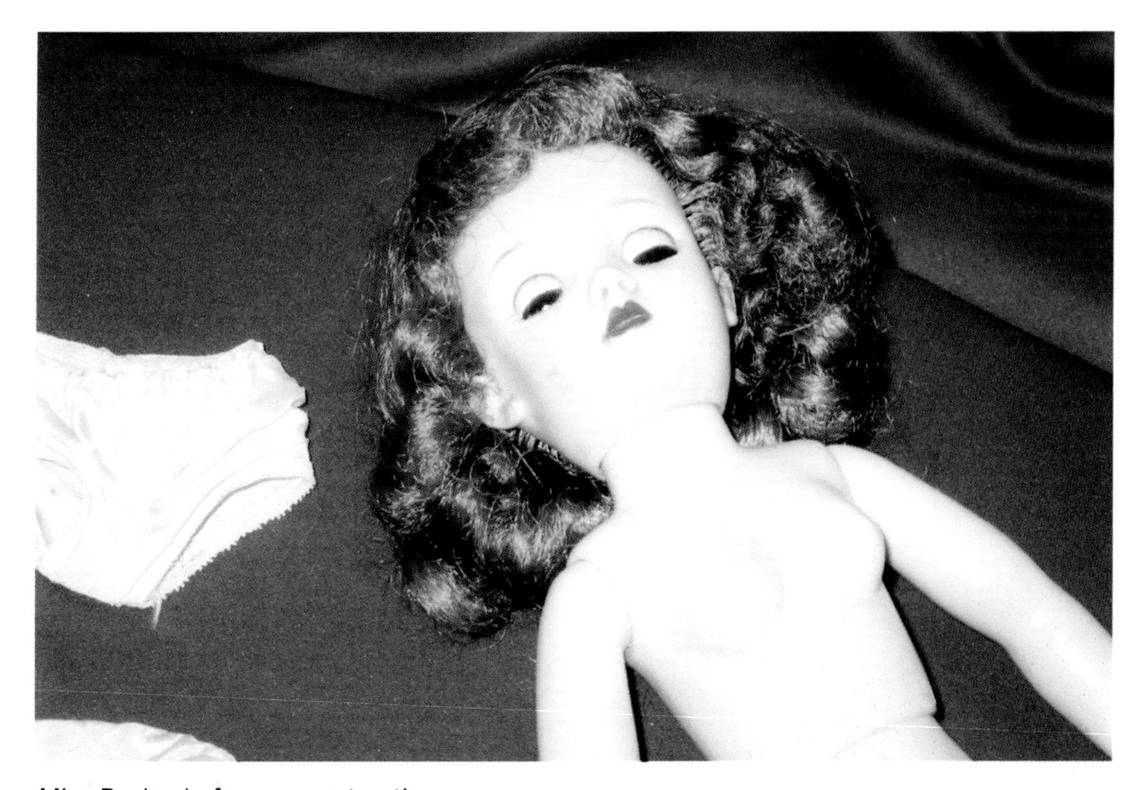

Miss Revlon before any restoration.

The original dress before restoration.

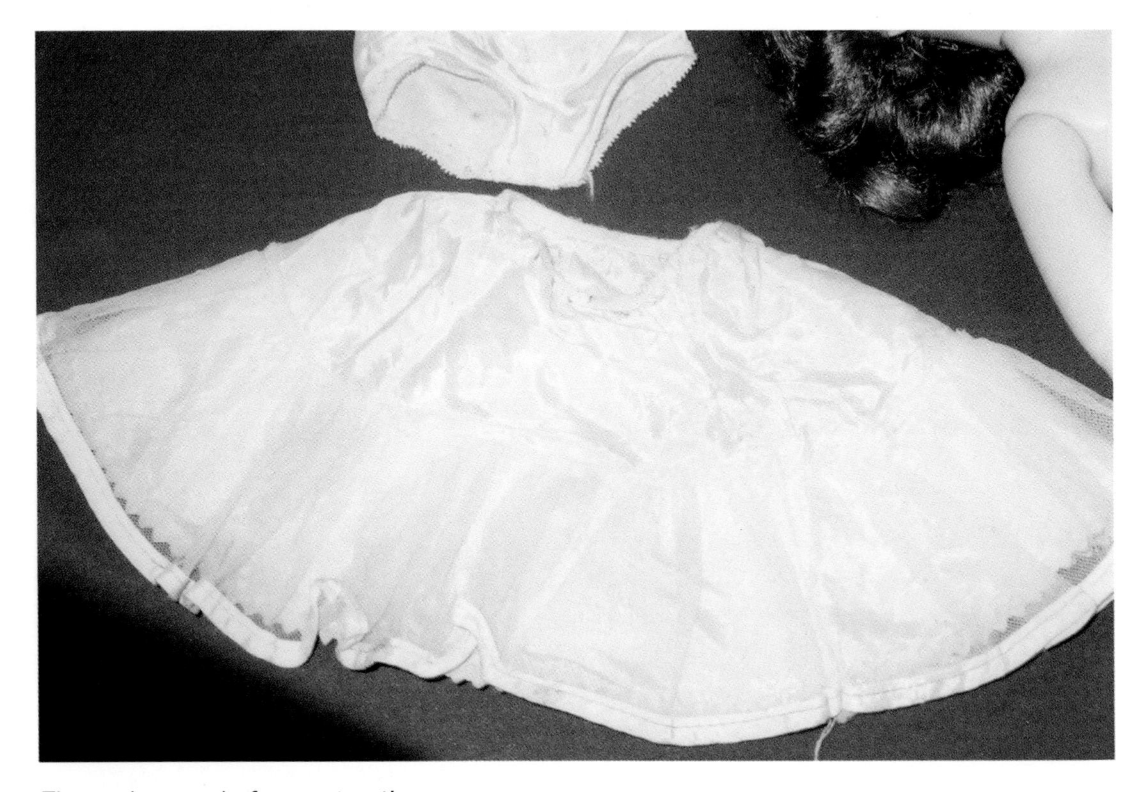

The underwear before restoration.

the condition of it (again, this is for your own protection, as you are bringing it to the customer's attention). Maybe she will say she can launder the clothing herself. That is fine but many will want you to handle that also. Mention what you will do such as sewing tears, adding a snap, laundering, ironing, fixing elastic in panties and whatever else needs to be done that you are willing to do. At the same time, you are summing up the work and what you are going to charge for it.

It is very important to list on the sales receipt **everything** and **every step** you are going to include in the restoration process along with the price you are going to charge.

The following are some pricing suggestions:

Restringing:	$ 15.00
Cleaning the entire doll	$ 15.00
Wash and set hair	$ 15.00
Treat the eyes	$ 5.00
Launder and mend the clothing*	
New elastic in panties, add snap to dress, fix rip in underarm seam	$ 12.00
Total:	**$ 62.00**

*Prices for laundering and mending the clothing will vary depending on the condition and the amount of work involved in laundering and mending the clothing.

If the customer wants to fix the clothing herself, be sure to cross out anything concerning the clothing and note that the customer took it with her.

If the customer needs the work done by a certain date, then make a **big note** on the receipt about that!

After you have written down on the sales receipt all the steps you are going to take to restore the customer's doll, and any other pertinent information, sign the receipt, giving the customer the original copy, and placing the second copy with the doll. This way you have a complete record of the doll itself, the clothing and accessories you have which belong to her, and the owner's name and contact information.

Leave the third copy of the repair order in the receipt book! The pages of the receipt books are numbered sequentially which will allow you to look up a customer's repair order should you need to do so.

In addition, the pages in this receipt book will be in chronological order, enabling you to find the date you received the doll. The second copy will stay with the doll until the work on the doll is completed and the owner picks it up. At that time you will note on that copy how you were paid — check, cash or other — as well as the date the doll was picked up. If your customer notifies you that someone else is picking up the doll for her, be sure to put the other person's name and telephone number on the receipt! Again, this is for your own protection.

Strive for a quick turnaround of the restoration. Almost nothing is more irritating for a customer than to have to wait for an extended period of time for the work to be done. If you have promised the work to be done by a certain date, make every effort to meet that deadline. If you find that you are going to run a little late, it would be courteous to call the customer and explain that you are going to be delayed.

If all the doll needs is a simple repair, such as restringing, try to get to that as soon as possible. Your customer will be delighted when you call and tell her that her doll is ready in a short period of time!

For the dolls that need more extensive restoration work, try to get to them in the order in which you received them, unless, of course, you have a specific date promised for any one of them.

Word about your restoration work will travel fast and soon you will be inundated with work!

TIP For a long time, I used shoeboxes for customers' dolls. I put the customer's name and the two dates on the end of the shoebox — first, the date the doll was brought in to be restored, and second, the date agreed upon for the restoration of the doll to be completed.

Restringing Dolls

Restringing a Doll

1 Measure a piece of round elastic approximately two-and-a-half times the length of the doll body from the shoulder to the hip.

2 Find the middle of the elastic and secure a hook to it. (SEE #2A.) The opposite end of the hook is placed inside the head of the doll and hooked to the rod attached from ear to ear. (SEE #2B.) (Some dolls will have a wire bale that is attached to the inside of the head above each ear. This bale swings freely and is easily seen at the neck. Secure the hook around this bale.) Be very careful **not** to secure any part of the hook or elastic on any part of the eye mechanism.

If there is no bale (like a handle on a bucket of paint) or rod in the head to which you can attach a hook, you will need to invent something to insert in the head onto which you can place the hook. I have used everything from a nail to a piece of a wooden dowel rod. Just use something that is long enough so that when you insert it into the head, you can lay it sideways on the inside of the cheeks. To make it easy, after you have determined the length you need, take it back out and clamp a hook, or even tie the elastic to it in the middle. You may wrap both sides of it with masking tape, so the elastic cannot slip off the ends.

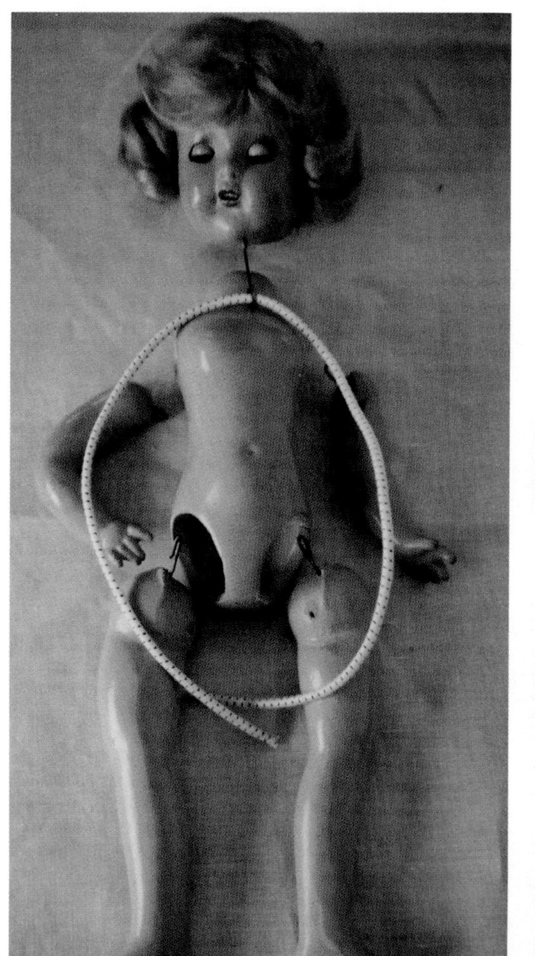

LEFT: STEP #2A.
Measure the elastic and find the middle for the hook.

BELOW: STEP #2B.
Place the hook around the rod in the head.

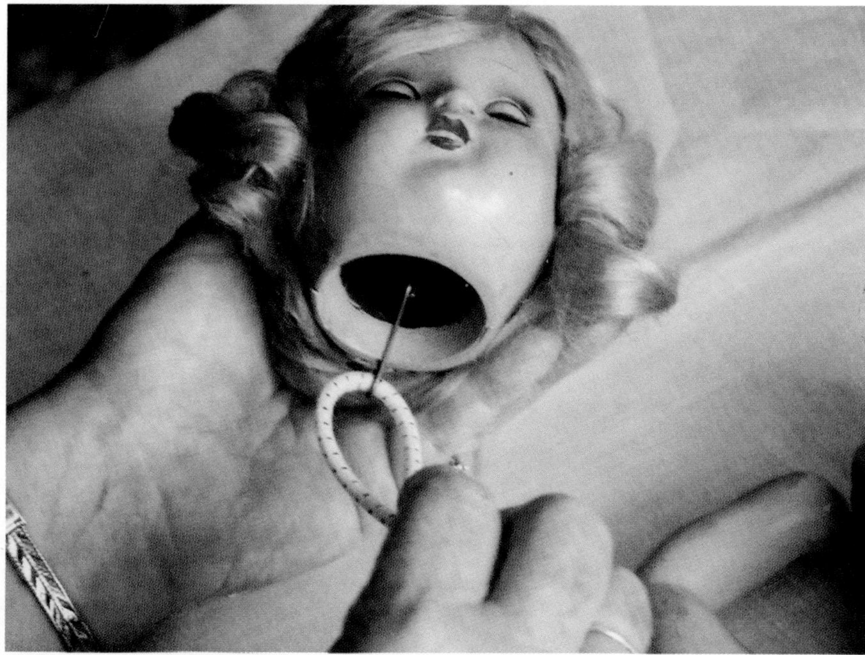

3 Feed each end of the elastic into each armhole, securing one end of the hook around the elastic and the other onto the arm. Repeat on the other side. (SEE #3A & 3B.)

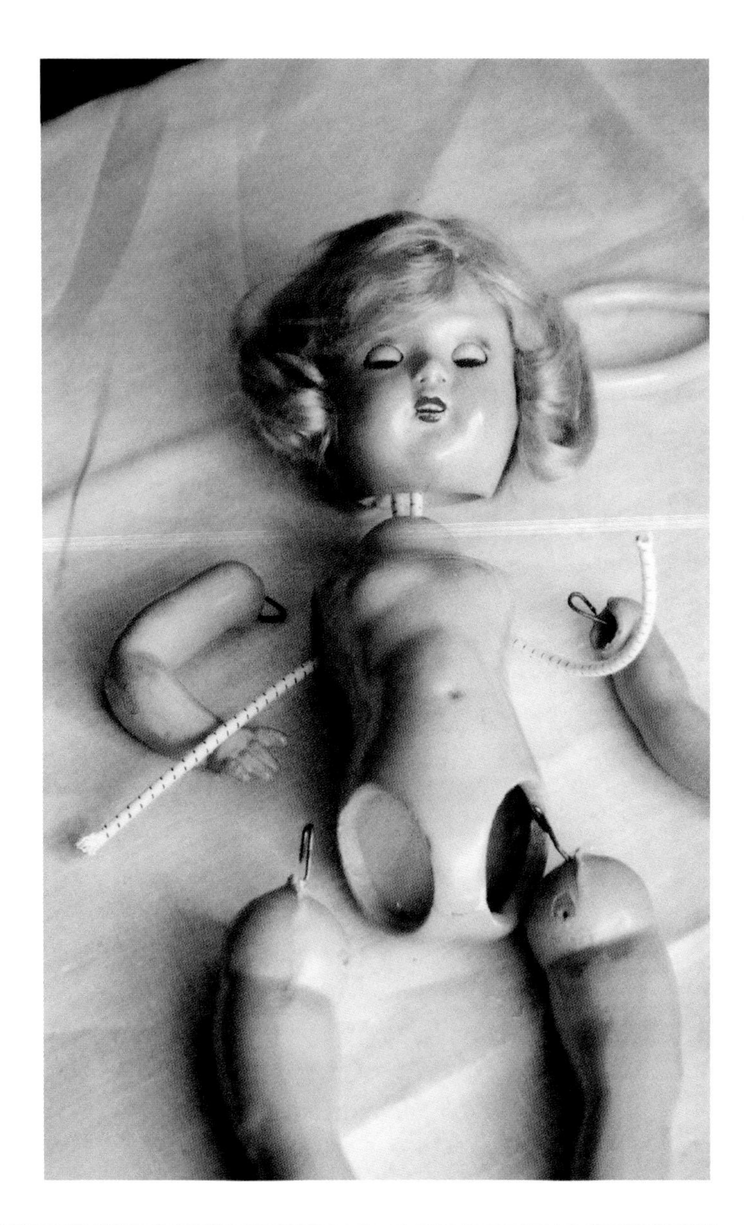

RIGHT: STEP #3A.
Feed the elastic through the armhole.

BELOW: STEP #3B.
Feed the elastic through the *other* armhole.

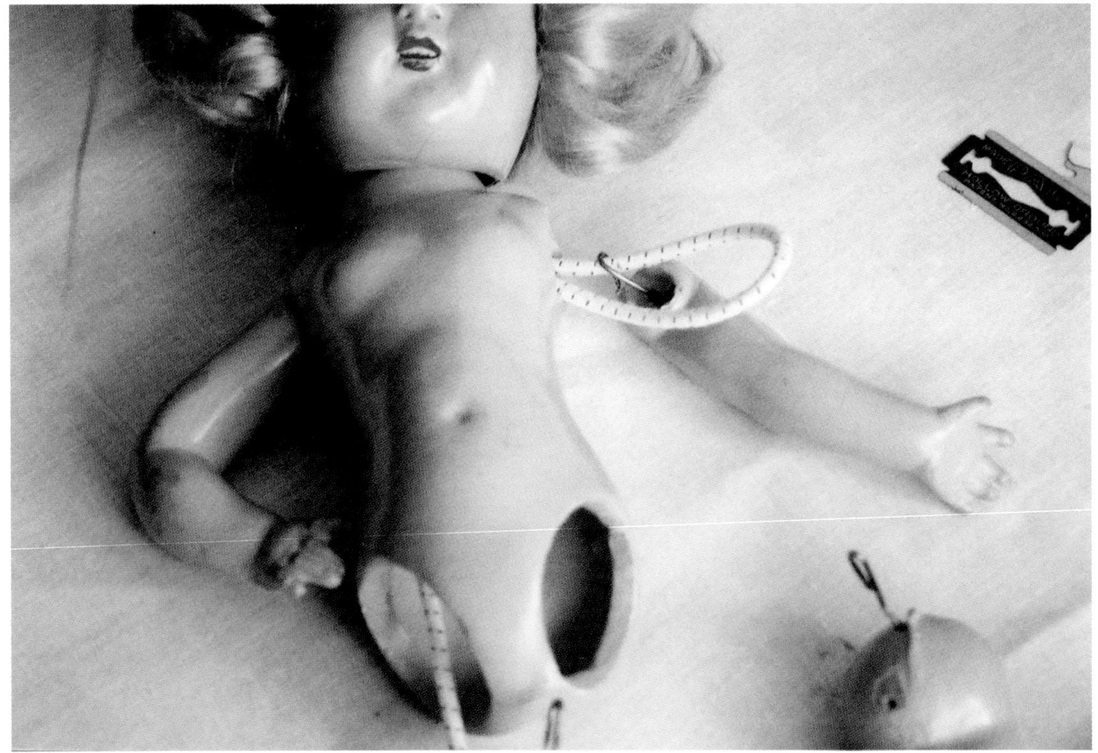

4 Feed the ends of the elastic through the body with the ends coming out of each leg hole. Secure the hook of the leg nearest you to the elastic. Then feed that same end of elastic through the crotch of the body to the other leg hole. (**SEE #4A.**) Secure the hook of the remaining leg to this same piece of elastic. (**SEE #4B.**) **This is important** — the knot in the elastic should end up between the arm-hole and the leg hole and **not** between the legs at the crotch, as the elastic will not hold the limbs as well.

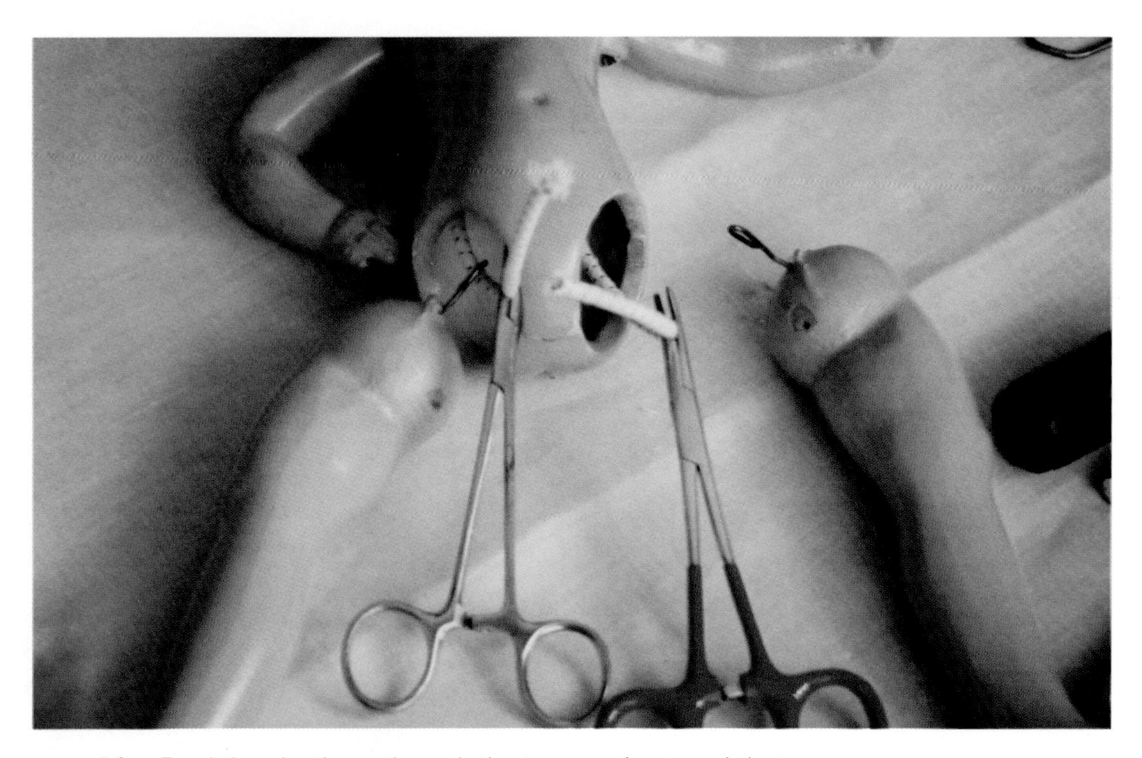

STEP #4A. Feed the elastic on through the torso and out each leg.

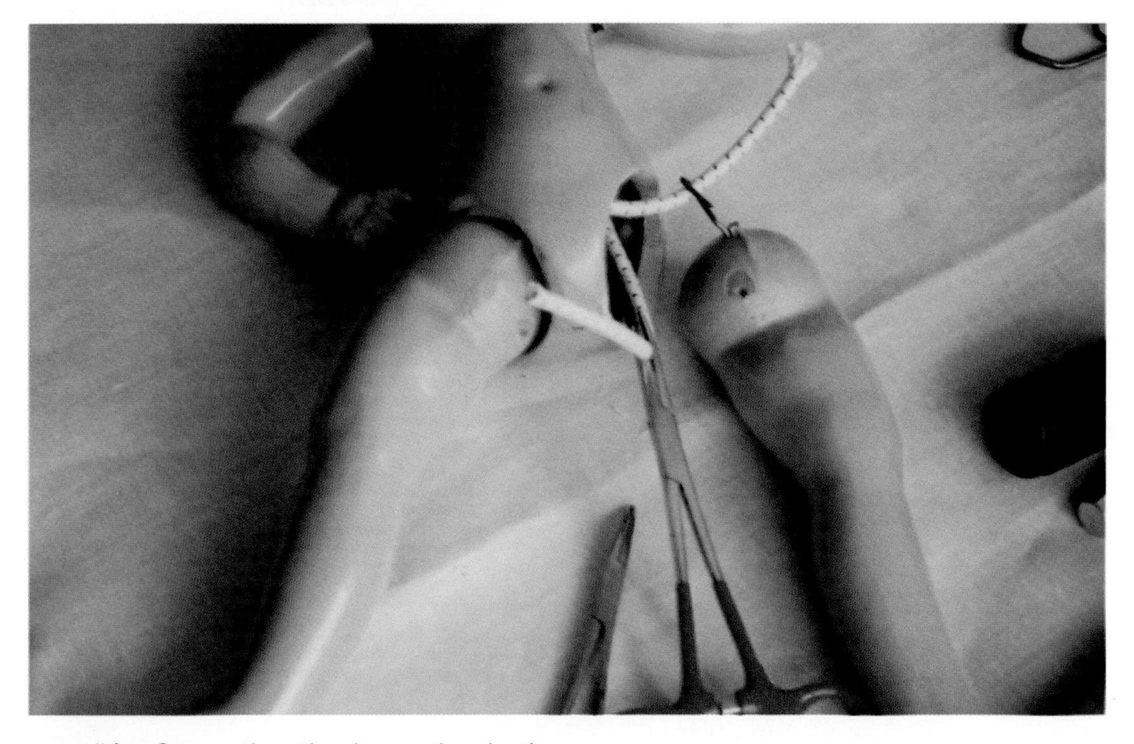

STEP #4B. Secure the other leg on the elastic.

5 **Pull the two ends of the elastic, stretching it tight.** Then tie the ends together at least three times. (SEE **#5.**) A surgical clamp is a wonderful tool to use at this time to hold the ends of elastic tight while you tie the knots. Cut off the excess elastic leaving about 1½ in (4cm) on each end.

6 This method is "one-piece elastic." (SEE **#6.**)

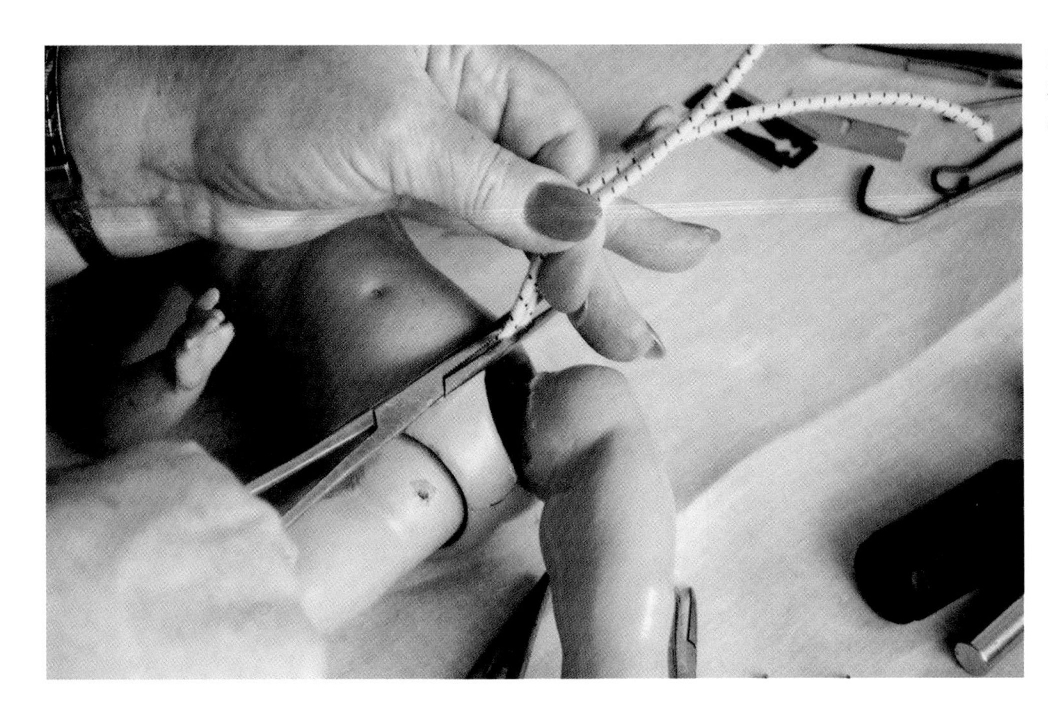

LEFT: STEP #5.
Stretch the elastic tight and clamp.

BELOW: STEP #6.
Diagram of restringing a doll using one-piece elastic

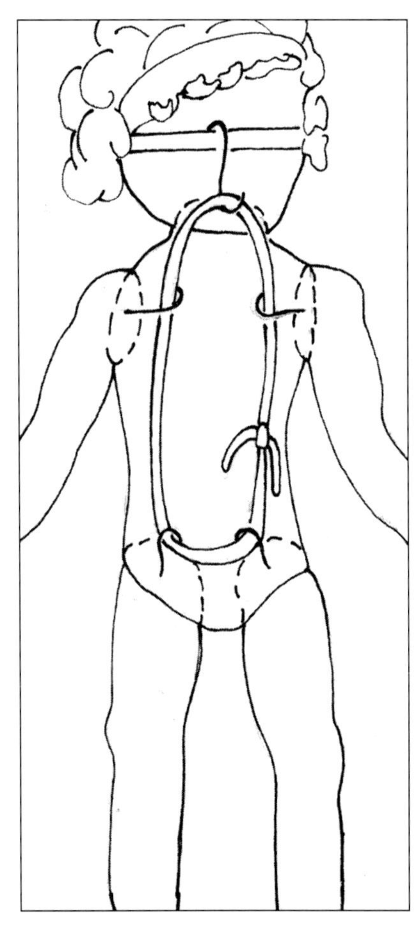

TIP When you are finished, if a limb wants to swing to the back rather than stay where you pose it, pull that arm or leg out of the socket, undo the hook and reverse the way you put the hook through the elastic. Slip a pencil through the elastic to hold the limb out of the socket while you change the direction of the hook.

TIP If you are going to string several dolls, or are starting a business, you will find the surgical clamp is a constant companion. It is handy for fishing the ends of the elastic down through the body. (I have several pairs of these in different sizes and lengths and would be lost without them now.) They can usually be purchased at a flea market where all kinds of scissors, dental instruments and other similar items are sold. They may also be available in hobby or craft stores as well.

Restringing the Head to the Legs, and Arms Separately

1. String the head to both legs. Using a piece of elastic that measures from one leg opening in the torso, up to the neck and back down to the other leg opening is recommended. It will have plenty of stretch to make it long enough to tie. It needs to be tight enough that the head and the legs can be posed when finished.

2. Divide the length of elastic into thirds. At about one-third of the length, run the hook through the center of the elastic so that you have two ends, one to each leg.

3. Put the head in place on the body, and reach up inside to get the ends down to each leg.

4. Run the elastic to one leg and hook it and then go to the second leg and hook it. Tie the knot between the head hook and final leg hook rather than between the legs, as the knot may not allow the legs to move properly if placed in the crotch.

5. To string the arms, use another piece of elastic. Run it through the hook on one arm, and put the arm in place on the body.

6. Reach through the second armhole and pull both ends of the elastic out.

7. Put one end through the hook on that arm and then pull tight and tie. If you have surgical hemostats, they work wonderfully to hold the elastic at several points while you are working. It is important to hold the elastic tight until you can get it tied.

8. If you are stringing an antique doll with ball joints, the arms are always strung separately from the head and legs. (SEE #8.)

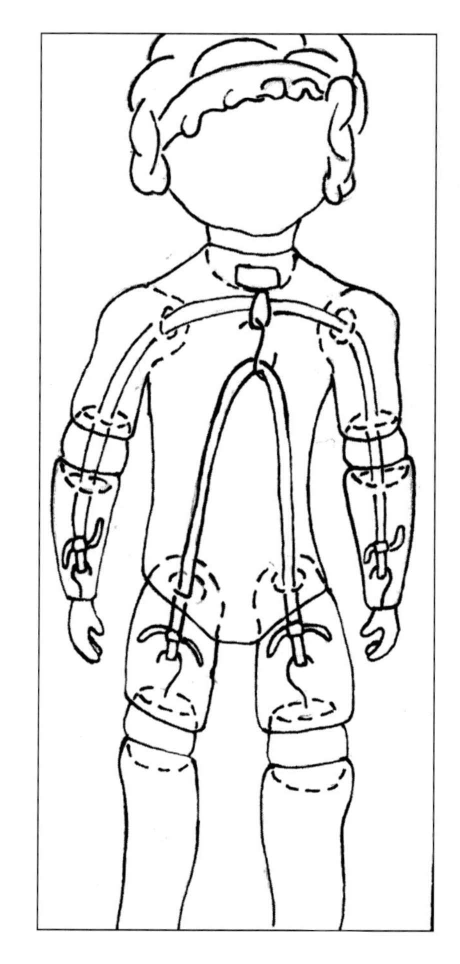

STEP #8. Diagram of restringing an antique ball-jointed doll

Restringing Using Hog Rings and Pliers

1 When you are stringing a large doll and using the heavy elastic, it is easier to use hog rings and pliers rather than tying the elastic when you are finished. Heavy elastic can be quite hard to hold tight and get it tied. Special pliers are made to use with hog rings. (SEE **#1.**) Insert the hog ring into the jaws of the pliers and work the two pieces of elastic into the ring.

STEP #1. Hog ring pliers and a hog ring in them.

2 When you have the elastic stretched as tight as you want it, apply this ring as close to the opening of the doll's leg as you can. Squeeze the pliers shut until the elastic is caught securely by the hog ring. (SEE **#2.**) This takes some practice to get it clamped tight.

3 Sometimes it is helpful after the first clamp of the pliers to move the jaws around the hog ring so that the pliers are actually over the edges of the hog ring and squeeze again. This will clamp the ends of the hog ring down tighter on the elastic.

4 If you are able to pose the doll's arms and legs and they stay in that pose, you have pulled the elastic tight enough. If they do not, cut the elastic and start over.

5 If you have not clamped the hog ring tight enough to hold the elastic, it will let the elastic slip back through the hog ring and the doll will not be tight. (This does take some practice but it is well worth it to learn to use these pliers and hog rings. You save elastic and it really is easier than tying the heavy elastic once you master the technique.)

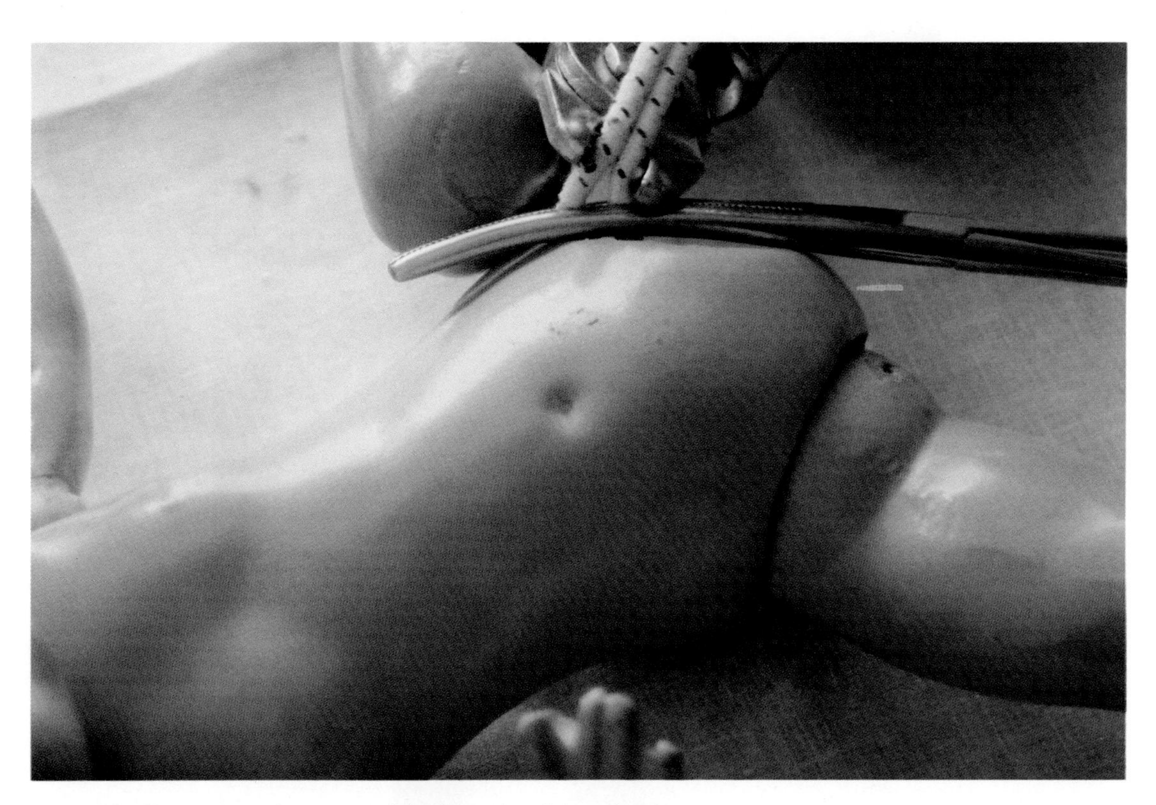

STEP #2. Clamp hog ring around **both** ends of the elastic.

6 Cut the ends of the elastic off about 2in (5cm) from the hog ring. (SEE **#6.**)

7 Let it all slip up into the torso of the doll.

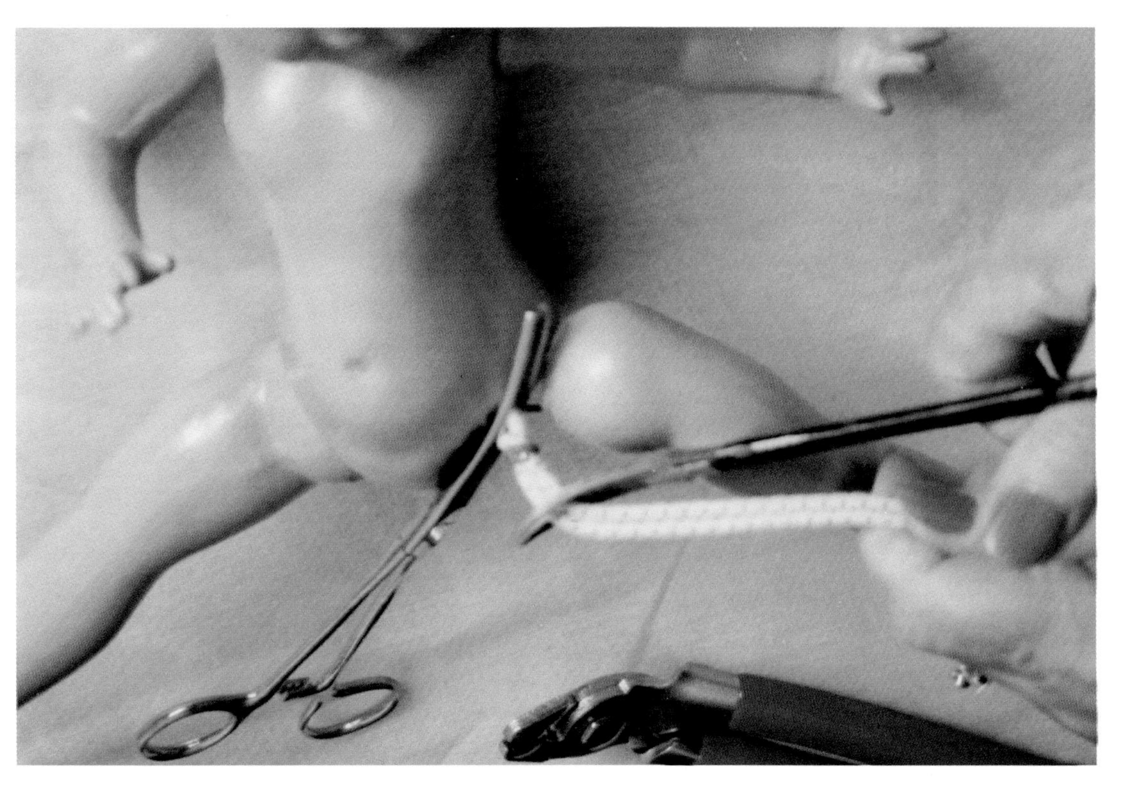

STEP #6. Cut off the ends of the elastic.

Restringing a Celluloid Doll

1 Celluloid dolls are not easy to restring. Very carefully remove the stretched-out elastic from the limbs. Pull out the knot that is inside the limb. It will come through the hole with a little persuasion. By holding your fingers over the hole, you can warm the celluloid and it will come out more easily. This takes time, but it works. Do this with all four limbs.

WARNING: Do not heat the celluloid with a hair dryer or near a flame of any kind. It will explode!!

2 The best elastic to use is sold on a card and called Elastic Cord. It comes in black and in white. It is not very heavy, but a celluloid doll is so lightweight that it does not need heavy elastic to hold it. This is not the "elastic thread" that you can put on your machine bobbin and sew. This is wound on a card and is similar to that which is used on a Halloween mask to hold it on your head. You can purchase it at any fabric store or department store.

3 Cut a piece of elastic about 6in (15cm) longer than the one you pulled out of the doll.

4 Tie a knot in the end like the one you pulled out of the doll, and place that knot right next to the hole in which you want to insert it.

5 Take a crochet hook or any small blunt item and gently push the knot through the hole into the limb. Again, you may need to warm the area with your hands. (A "blunt" item is recommended because if you try to use a large pin or large needle to push the knot into the hole, it will pierce the elastic and will not push it through. Also, be sure your blunt item is not bigger than the knot. You do not want to make the hole in the celluloid any bigger because you want your knot to hold the elastic inside.)

6 When you have secured one in a limb — say one leg — then run the elastic through the doll's crotch and out the side for the other leg.

7 Pull the elastic tight and clamp it with a hemostat so that it will not go back into the body.

8 Tie a knot in the elastic near the hemostat pliers, leaving enough room to be able to work. Push that knot into the second leg. (After you have tied the second knot and are ready to push it into the second leg, cut the elastic off about 1in [3cm] or so beyond the knot.)

9 Force this knot into the hole in the second leg. Hopefully, you will have stretched the elastic tight enough that the legs will be held close to the body and yet be movable so the doll can be posed. This is why you clamp it off after stretching the elastic, so it will snap back when the second leg is strung and will hold them correctly. If not, start over — now that you know what you need to do to make them fit properly.

10 Use this same technique for the arms.

TIP If, by chance, the celluloid has split so much that the hole is bigger than it should be, you can use the adhesive strips from a thin Band-Aid®. Cut the strips off; do not use the part with gauze. Cover the area with the adhesive strip and make a new hole through it. The Band-Aid® adhesive strip will stick to the celluloid nicely and give it some reinforcement. Also, it will not show when the limbs are strung, as it will be hidden in the joint area.

Restringing a Little Miss Revlon Doll

1 Hold the leg joint of the doll under hot water for a couple of minutes and then gently twist and carefully pull the leg off to expose the hip cavity. Inside you will find the cotter pin, a metal washer and a bell-shaped piece of plastic. Remove the pin and washer and leave the bell inside. If the pin is missing, make a new one by cutting 1in (3cm) from a large paper clip or bobby pin with tiny pliers and bending it into shape.

2 Put the rubber band/elastic loop on the bobby pin so the pin will act as a needle for threading parts onto the band/elastic loop.

3 Put the bobby pin through the metal washer and catch the band/elastic loop with the cotter pin.

4 Put the bobby pin inside the hip piece and into the bottom of the bell piece, out the top of the bell piece and out the top of the hip piece.

5 The rubber/elastic loop band is probably stretching quite a bit by now, so grasp the bobby pin at the top of the hip piece with the pliers and pull the assembly tight.

6 Put the bobby pin into the flat side of the tube guide and out the rounded end.

7 Put the bobby pin into the waist chest piece and out the neck. Again, grasp with the pliers and pull tight.

8 Push the bobby pin through the neck collar and pull up so the bobby pin is all the way out of the doll and the rubber band on it is exposed.

9 Put the second cotter pin on the rubber band to hold the whole assembly together and slide the bobby pin off the rubber band.

10 You should now have a doll with one leg and the head missing. Twist the head onto the neck collar and the leg back into its socket and it is done!

Restringing Very Small Dolls (Such as All-Bisque Dolls)

1 Small dolls like those shown in the illustration (SEE **#1.**) are easiest to restring with the round elastic cord that is sometimes used in dressmaking. You can purchase this in the sewing notion department at any fabric store. It usually comes on a card. If the head is stationary on the torso, there are only the four limbs to string.

2 Run the elastic from one leg to the opposite arm and then continue on and run it to the other leg and arm, making a crisscross inside the body.

3 Tie it off. If, when you finish, one leg or arm wants to turn backwards, hold the doll up and run the elastic through that particular hole in that limb in the other direction.

4 If a doll has a neck joint, string the head to the two legs. String the arms across to each other (with separate stringing). Do not string two legs together because they are so close in the crotch that they will just dangle, so use something up higher in the doll such as the arms. Sometimes a doll will have a bar across the inside of her torso that you can loop the elastic over and string the legs.

STEP #1. Small bisque dolls.

5 When the doll only has joints at the shoulders, all you need to do is string the arms straight across from one another. That type of doll usually has openings inside her arms at the shoulders to run the elastic through. (SEE #5A.)

If that area has been broken and you have no hole for the elastic, you can repair this area with acrylic fingernail materials which may be purchased at a local beauty shop supply store. (SEE #5B.) Put a bead of the acrylic powder mixed with the liquid onto the damaged area. Use your imagination a little and make a hole in the mixture large enough for the elastic to go through easily before it sets up! A wooden toothpick will do the job. Take it out in time for the acrylic to finish hardening. You can even use an emery board to clean the edges a little if needed! (For instructions on using the acrylic fingernail materials for making replacement fingers for composition dolls, see the chapter on "Composition Dolls.")

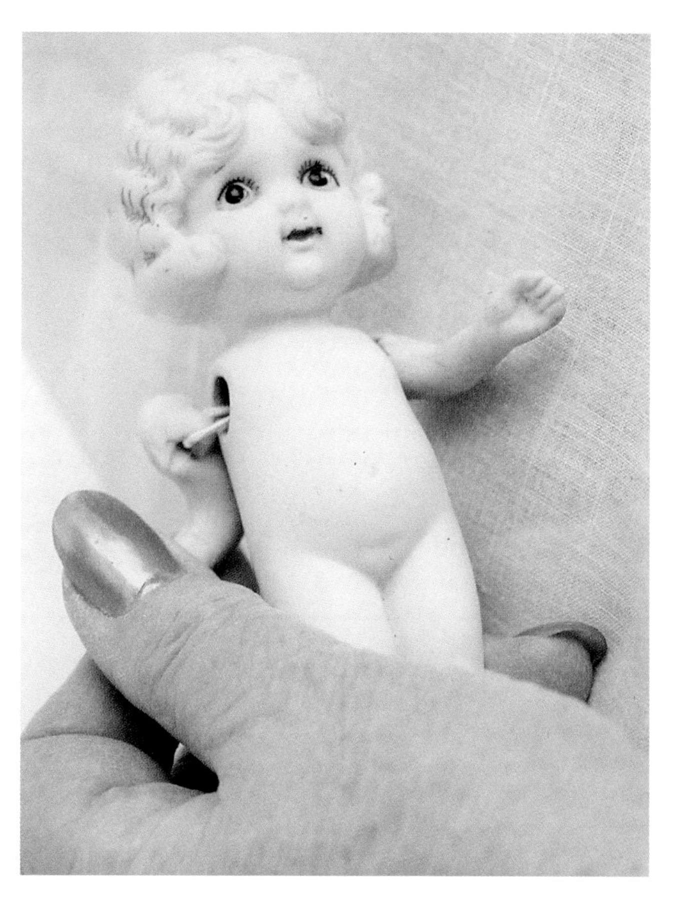

LEFT: STEP #5a.
Jointed at the shoulders only and strung with elastic.

BELOW: STEP #5b.
Nail acrylic materials.

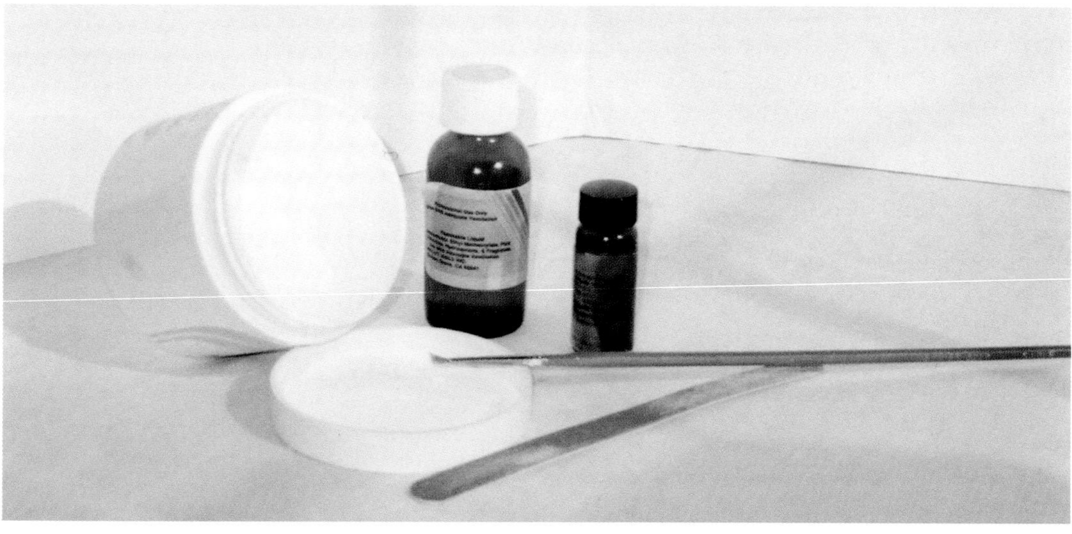

Restringing Little Dolls with Wire or Elastic

1 The first clue that you need to string them with wire is that they will have a hole all the way through the arms and legs so that you can run the wire completely through them. (SEE **#1.**) If they do not have these holes, they should be restrung with hooks and elastic.

2 Usually the arms are wired straight across from one another as are the legs. Use your needle nose pliers to twist a little circle at the end of one piece of wire.

3 Run the other straight end through the limb, through the doll and out through the other limb. Leave the wire long for now and use your pliers to make the circle, which will fit against this limb like the first one. When you have it bent enough to see how much wire it will take, then cut the wire off and finish making the circle tightly against the limb. This takes some practice.

You may see small dolls like this with elastic. (If you are using elastic, you will need very fine round elastic for the project.) Sometimes the elastic is merely tied in a knot so that it works like the circle of wire and will not pull through the hole in the limb. Then, of course, you will need to stretch it tight when tying the knot on the outside of the opposite limb. Elastic is not the proper material for all-bisque dolls, but if you are going to dress them and cover their joints, it will be acceptable for your own dolls.

STEP #1. All-bisque doll strung with wire.

Cleaning Dolls

Over the years, I used many kinds of cleaners to restore dolls — both my own and those for customers. Every time I saw a new product on the market, I purchased it and tried it. As a result, I have developed my own cleaner — Carol's Miracle Doll Cleaner — which will clean any kind of doll. (SEE PHOTO ON THE TOP RIGHT.)

Carol's Miracle Doll Cleaner.

CAUTION!

Before you begin your cleaning work, be sure to read all of the instructions first and follow the recommended use.

It is a good idea to dip a cotton swab in Carol's Miracle Doll Cleaner and rub it on the doll in an inconspicuous area to be certain it does not damage the doll. This procedure demonstrates improvements to the doll and protects the doll from unlikely damage should there be an adverse chemical reaction.

Please note that the features on some of the hard plastic and composition dolls are painted on and, therefore, not permanent. Others are permanent, but you cannot always tell right away which are which. (SEE PHOTOGRAPH ON THE BOTTOM RIGHT.)

To test, dip a toothpick in the cleaner for just a tiny dab of cleaner and barely touch the end of an eyelash or eyebrow. If the paint comes off, you will know the features are not permanent. Vinyl dolls and antique bisque dolls are usually no problem because the features on them are, in most cases, permanent. It is always a good idea to test first, however.

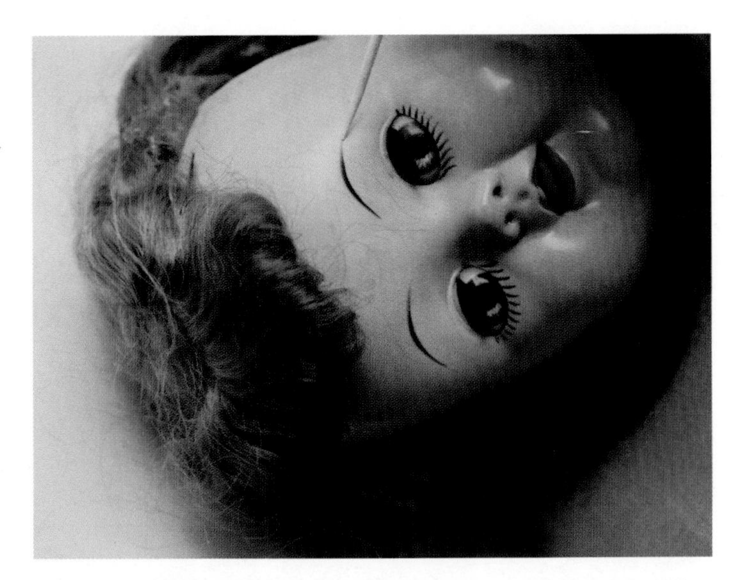

Test the features of the doll. Are they permanent or not?

Cleaning a Doll

1. Use an old washcloth – the rough kind, not the velvety ones.
2. Wet a circle on it with the cleaner. (SEE **#2.**)
3. With your finger on the washcloth in that area, start gently rubbing the doll. (SEE **#3.**)
4. Add more cleaner to that same area to continue cleaning.

STEP #2. Applying cleaner to the washcloth.

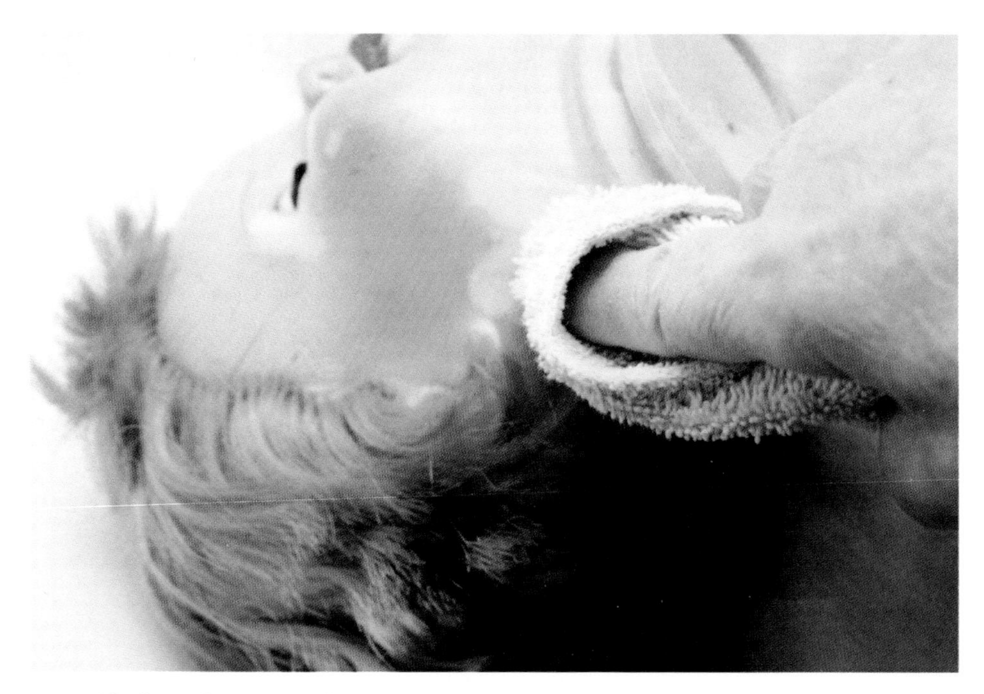

STEP #3. Cleaning the doll.

5 Put some cleaner on a clean place on the washcloth and go all over the doll again.

6 With a very wet cloth, clean the fingers, toes, eyes and limbs — wherever there are creases — leaving a little cleaner behind in the wrinkles and creases. (SEE **#6.**)

7 Use a wooden toothpick to clean the wrinkles and creases.

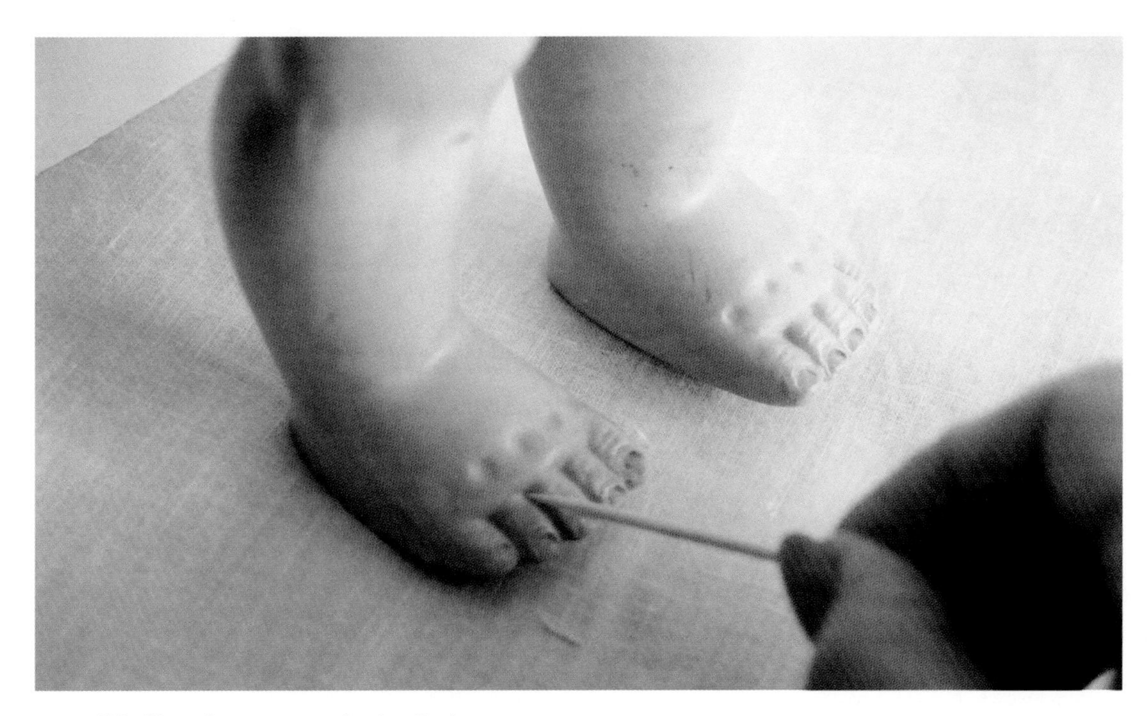

STEP #6. Cleaning grooves in the limbs.

Cleaning a Doll's Eyes

1 Many times, a doll's eyes are full of a white substance, which, in most instances, is soap scum from all the baths she had when her little mother played with her.

2 Lay her on your lap and hold her eyes open with one hand.

3 Dip a cotton swab in the cleaner and saturate her eyes with it. Do this while they are open and then again when you allow them to close.

4 Use a wooden toothpick (plastic ones might scratch) and clean out all the film. Her eyes will be bright again. This is good on acrylic, vinyl, tin and glass eyes.

Cleaning Mildew from a Doll

1 Put some of the cleaner on an old rough washcloth and rub it on over the entire doll, paying attention to wrinkles and creases in her fingers, toes and limbs. Allow some extra cleaner to remain in those wrinkles and creases while you are cleaning the rest of the doll.

2 Use a toothpick and rub through the wrinkles and creases, taking all the mildew out.

3 This process will probably take several hours, as you need to let the cleaner really cut the whitish film off the vinyl. Continue to put cleaner on your washcloth, rub some more and use the toothpick also.

Cleaning a Bad Smelling Hard Plastic Doll

1. First cut the elastic and work with the body in pieces.

2. Clean each part really well inside and out. (The best tool to use to clean the inside of the doll is a bottlebrush).

3. Stuff each piece – head, arms, legs and torso – with dryer sheets. (Any brand will do.) (SEE #3.)

4. Put all the parts into a plastic bag with tissue paper over the face and close it up. (SEE #4.) Caution: make sure the parts are completely dry before putting them in a plastic bag. If they are the least bit damp, a mildew or mold problem could occur.

5. Check on the doll after a week or so. Then take the doll out and remove the dryer sheets.

6. Let the doll lay out for a few days to see if the odor has been removed.

7. If the odor is not completely gone, stuff the doll again and repeat steps 3, 4 and 5. You may have to do this several times, but it will diminish the odor a little at a time, until it is gone!

8. If this is a doll you are going to display in your home, do not put her in a closed area with other dolls, such as a glass display case. If the smell should come back, everything in the case will pick up the smell. If the doll begins to smell again, repeat steps 3, 4 and 5.

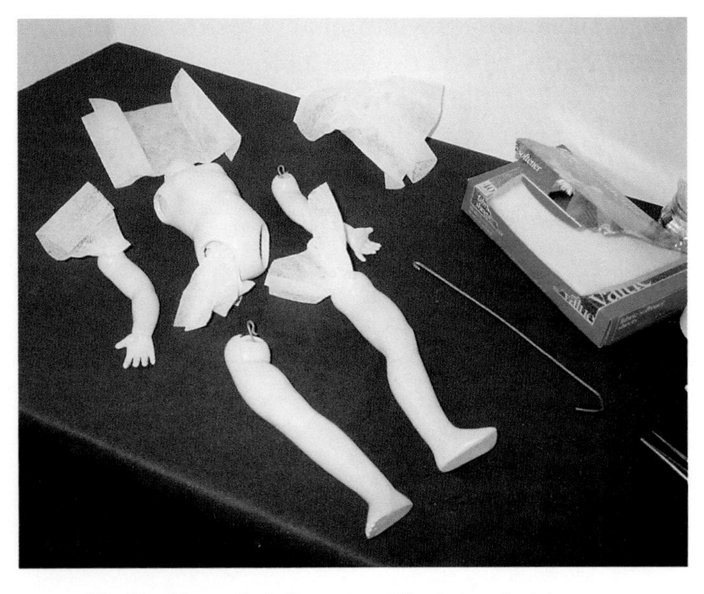

STEP #3. Stuffing all doll parts with dryer sheets.

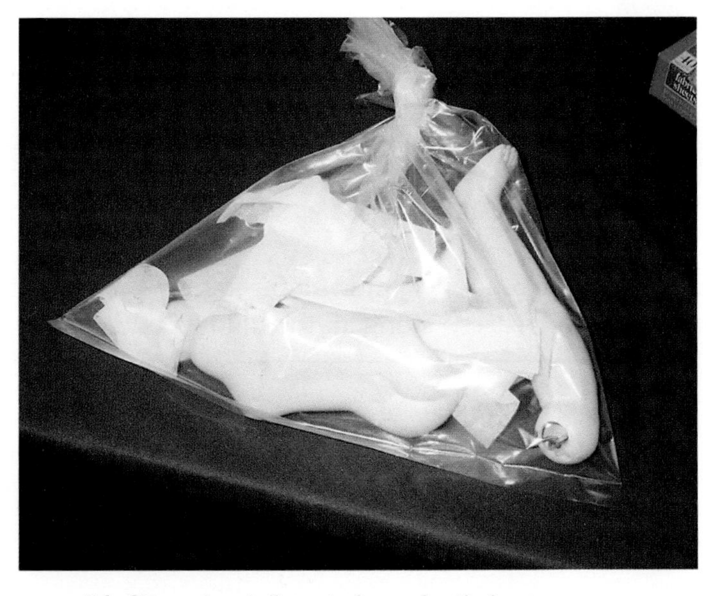

STEP #4. Store dry doll parts in a plastic bag.

Cleaning a Sticky and Gummy Doll

1 Put some of the cleaner on an old rough washcloth and rub it on over the entire doll, paying attention to the wrinkles and creases in her fingers, toes and limbs. Allow some extra cleaner to remain in those wrinkles and creases while you clean the rest of the doll.

2 Use a toothpick to rub through the wrinkles and creases, taking the dirt out.

3 Sprinkle your hands with baby powder or talcum powder and rub over the entire doll. This will make her nice and smooth again. (SEE #3.) Rub the powder in and get rid of the excess with your hands. You will not see the powder on her when you finish.

NOTE You might want to warn the doll's owner that the doll may become sticky again – depending on the contact she has with various substances. But the doll shouldn't become sticky for a long time after this cleaning process. If the doll does get sticky again, her owner can simply repeat step # 3, and the doll will be fine once again.

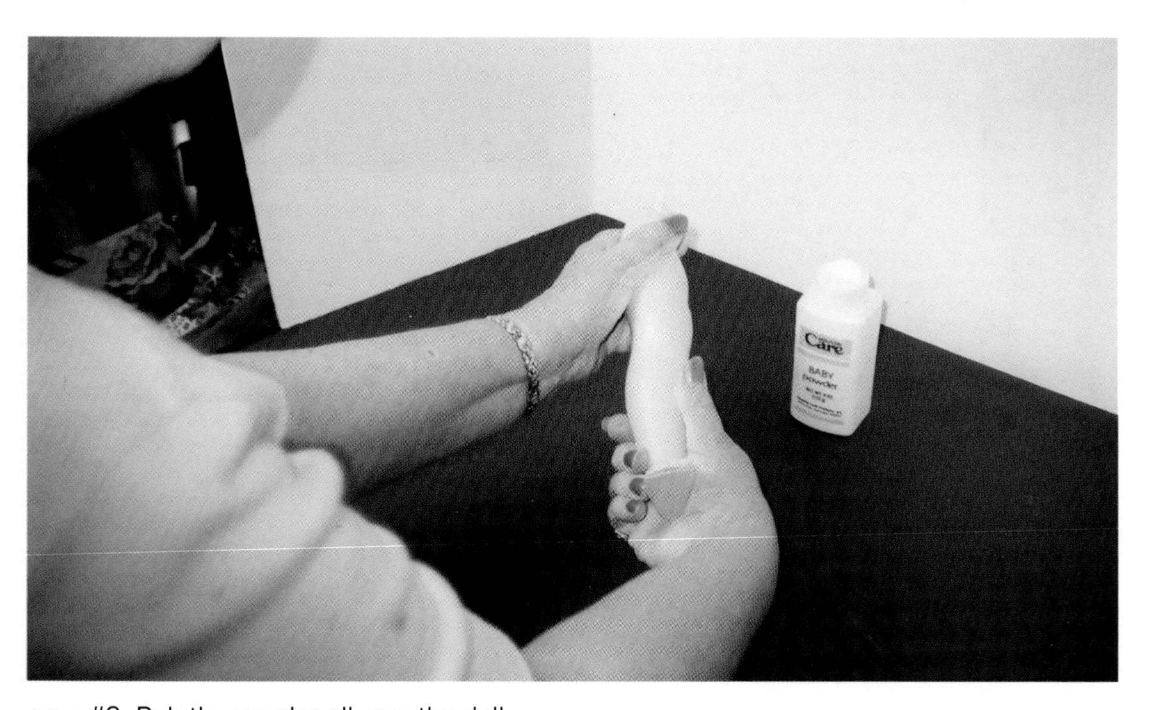

STEP #3. Rub the powder all over the doll.

Cleaning a Doll After a Fire

Cleaning a doll which has been through a fire and which has soot on it is no different from cleaning dirt off a doll. However, you may use a few more old washcloths and a little more cleaner due to the residue. In addition, the clothing and hair will need to be cleaned as well as the doll itself.

1. Completely undress the doll.
2. Clean the doll all over really well. (SEE **#2.**)
3. Using an old washcloth or a paper towel with a rough texture, wet a circle on it with the cleaner. (SEE **#3.**)
4. With your finger on the washcloth (or paper towel) in that area, start gently rubbing the doll. (SEE **#4.**)
5. Clean all the large areas of the doll first.
6. Clean the detailed areas, such as her fingers, toes and eyes. (SEE **#6.**)

STEP #2. Carol's Miracle Doll Cleaner.

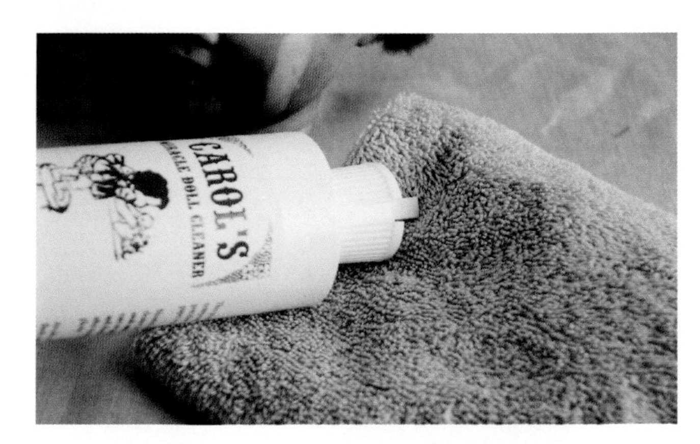

STEP #3. Applying cleaner to the washcloth.

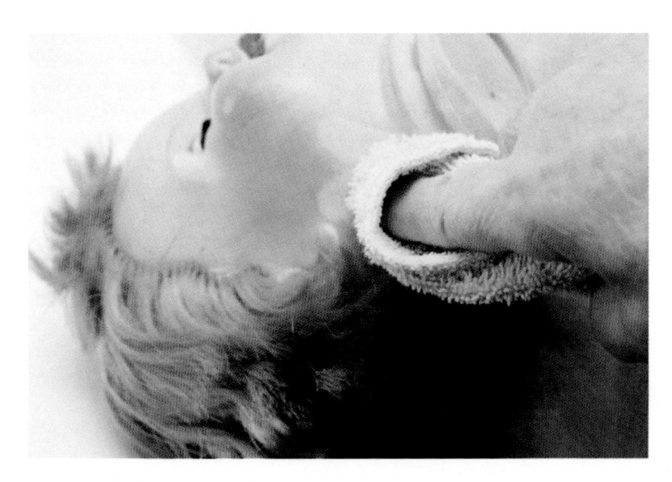
STEP #4. Cleaning the doll.

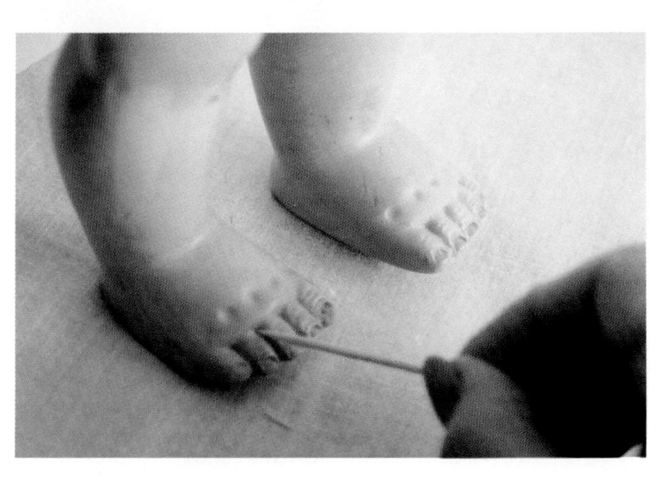
STEP #6. Cleaning grooves in the limbs.

7 Use a cotton swab and a wooden toothpick to get in the wrinkles and creases.

8 Shampoo the doll's hair. (For instructions, see the appropriate section on washing a doll's hair in the chapter on "Doll Hair.")

9 Launder the doll's clothing. (For instructions, see the section on laundering doll clothing in the chapter on "Doll Clothing.") If the clothing cannot be laundered, it may have to be dry-cleaned.

10 The doll's shoes can usually be wiped clean with the cleaner, even the suede ones! (For instructions, see the section on restoring doll shoes in the chapter on "Doll Clothing.") (SEE **#10.**)

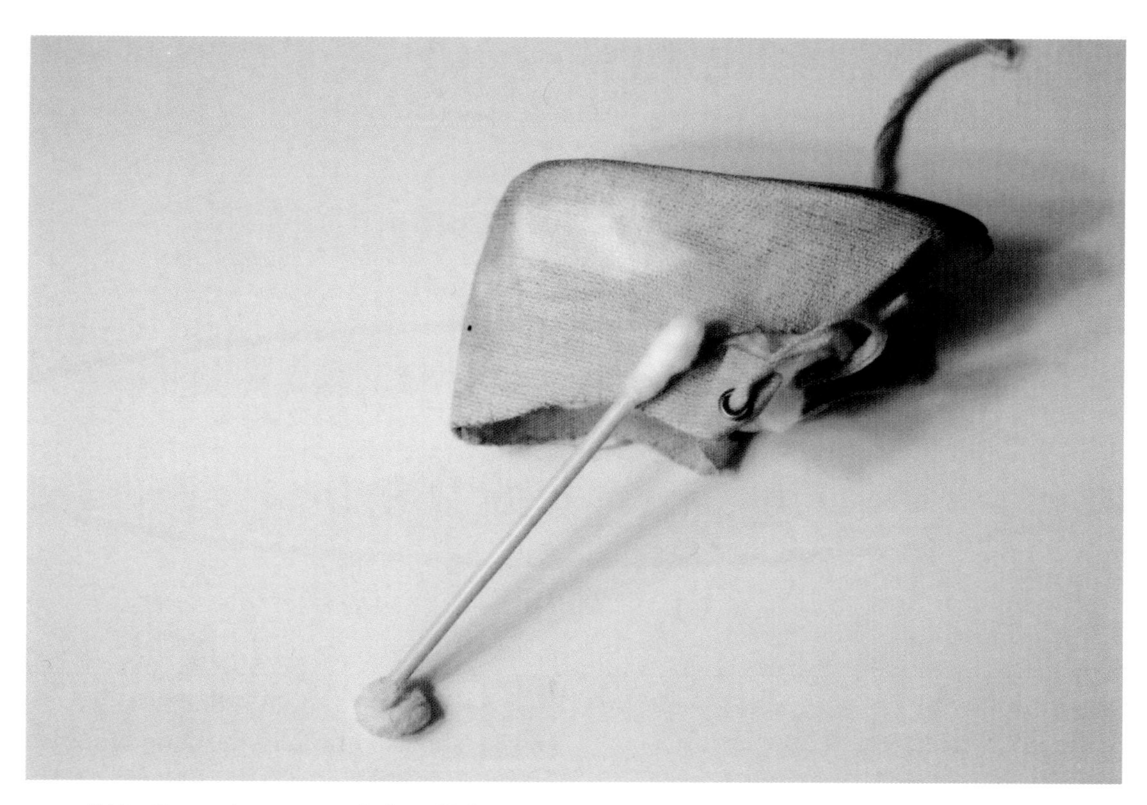

STEP #10. Clean the soot and dirt off the shoes

Cleaning a Doll from a Smoking Environment

If you have purchased a doll which has been in a smoking environment, you will be able to tell right away. The doll is salvageable, however.

To clean the doll, follow the previously detailed instructions in the section on "Cleaning a Doll After a Fire" in this chapter. After you have completed all the steps, you may pin some dryer sheets under her clothing and possibly even stuff some inside her torso, if needed.

Removing Ink, Marker and Green Ear Stains

1. Cover your doll with a paper towel except for where the ink or green mark appears.

2. Dab Oxy 10® (it is recommended that you use this product, not a substitute) on the ink, marker or green stain and place the doll in direct sunlight. The number of days your doll will be in the sun will be determined by how dark the spot is.

3. Each day, add a little more Oxy 10® on the spot. Eventually the stain will bleach out.

4. Patience is important here as it could take several weeks for the spot(s) to completely bleach out so do not be in a hurry!

Doll Hair

Washing and Styling the Hair on a Doll with a Wig Cap

The following procedures will work on both synthetic and human hair wig caps.

1 Hold the doll's head under warm slow-running water in the sink to wet it. Use liquid fabric softener for shampoo which will leave the hair clean and soft. (If the wig cap is on a composition doll's head, you need to be extremely careful about getting water on the composition as it can damage the composition.)

2 Rub the hair gently and rinse it out under the running water. Make sure all the fabric softener shampoo is rinsed completely out. The wig cap will probably get wet, but you can shape it when you are finished with the washing and rinsing process.

3 Wrap the doll's head in a towel. Pat and squeeze to get it as dry as possible.

4 Use the handle of a comb (a pick is fine) to straighten out the hair as best as you can while it is wet.

5 Now inspect the fabric wig cap. If the glue has loosened in any spots, pull the wig down around her head like it should be. (SEE #5.) Try to smooth the crown of the hair so that it will look nice.

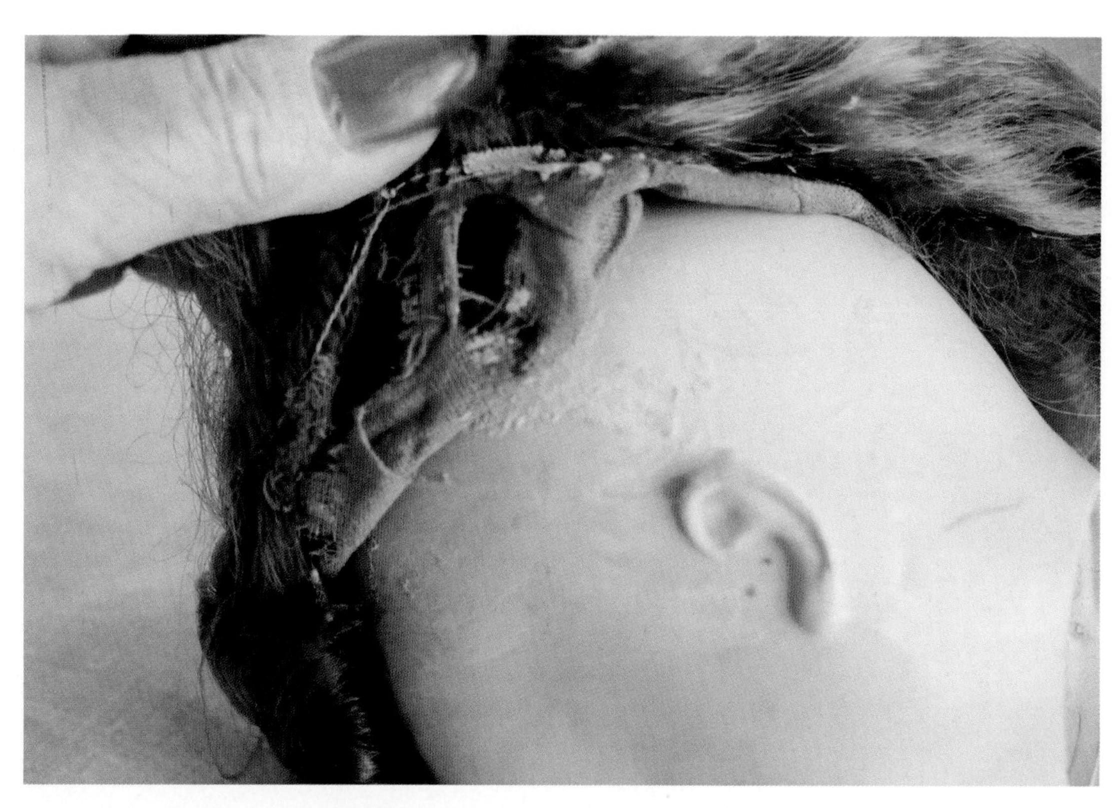

STEP #5. A wig cap on an old wig.

6 Let it all dry. (SEE #6A & #6B.)

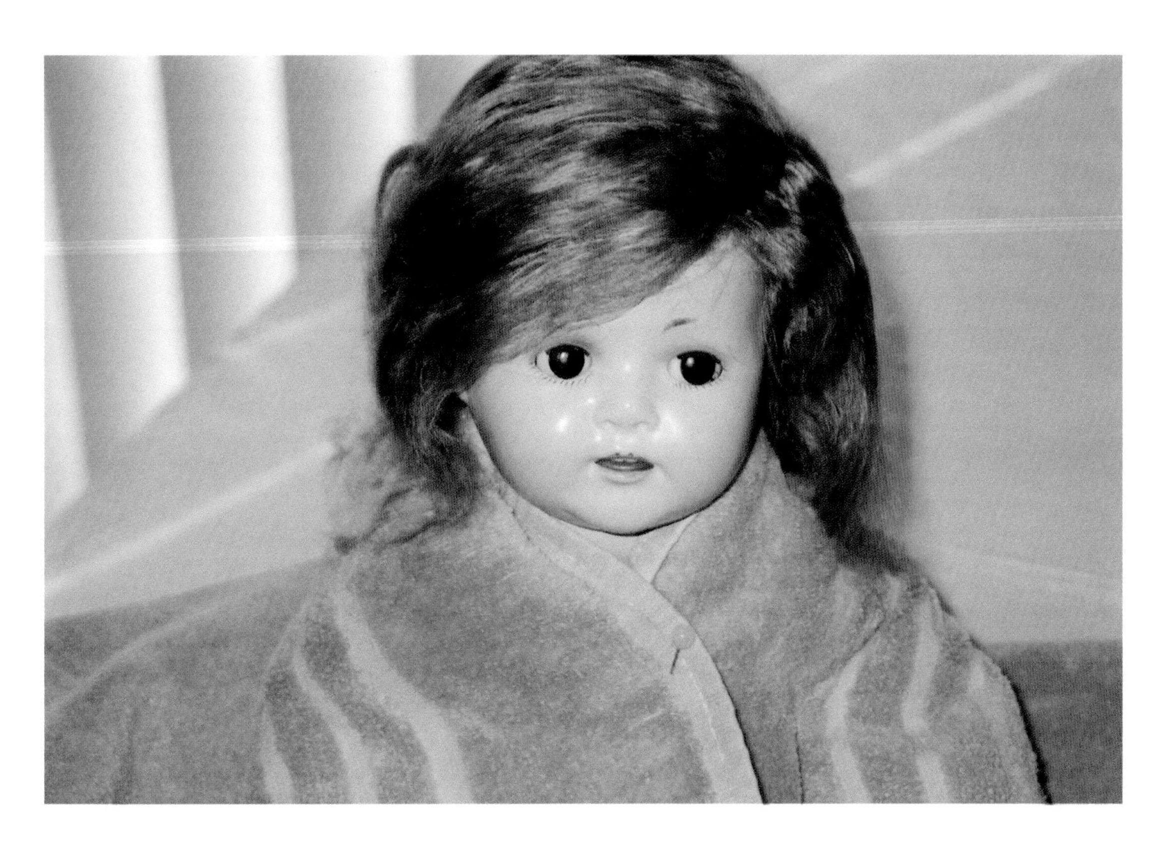

STEPS #6A. & #6B.Waiting for the doll's hair to dry.

7 Apply some glue between the head and the cap. Position the cap back in place. Let the glue dry thoroughly.

8 If the cap did not loosen, go ahead and fix the hair while it is wet. Otherwise you will need to wet it again a little at a time using a brush or comb dipped in water.

9 Home permanent curlers and papers can be used to set the doll's hair. Roll the curlers so that they go around her head in a row. When it dries and you take it down, it will look like long curls.

10 Let the hair dry at least twenty-four hours or more before removing the curlers. (SEE #10.)

11 When the hair is completely dry, unwind each curler carefully taking the paper off the ends and gently guide the curl back up into place.

12 The hair will relax some over time and the curls will fall and get longer. If you want to keep the curls tight, put a hair net on her.

STEP #10. Let it dry twenty-four hours or more.

Your doll now has clean and shiny hair in a new set!

She now has clean and shiny hair in a new set!

Washing and Styling Rooted Hair

1 Hold the doll's head under warm slow-running water in the sink to wet it. Use liquid fabric softener for shampoo which will leave the hair clean and soft.

2 Rub the hair gently and rinse it out under the running water. Make sure all the fabric softener shampoo is rinsed completely out. (SEE #2A, #2B, #2C, & #2D.)

3 Brush the hair and set it while it is wet.

4 Home permanent curlers and papers can be used to set the doll's hair. Roll the curlers so that they go around her head in a row. Or you may decide to braid the hair or set it in another style.

5 Let the hair dry at least twenty-four hours or more before removing the curlers.

6 When the hair is completely dry, unwind each curler, carefully taking the paper off the ends, and style it.

NOTE Before setting the doll's hair, you may want to do some research to find out how it looked when new and then try to style the hair as it appeared originally. However, if a curling iron has been used previously on the hair, it may be "fried" and, therefore, will not come out as well.

STEP #2A. Shampoo — Fabric Softener.

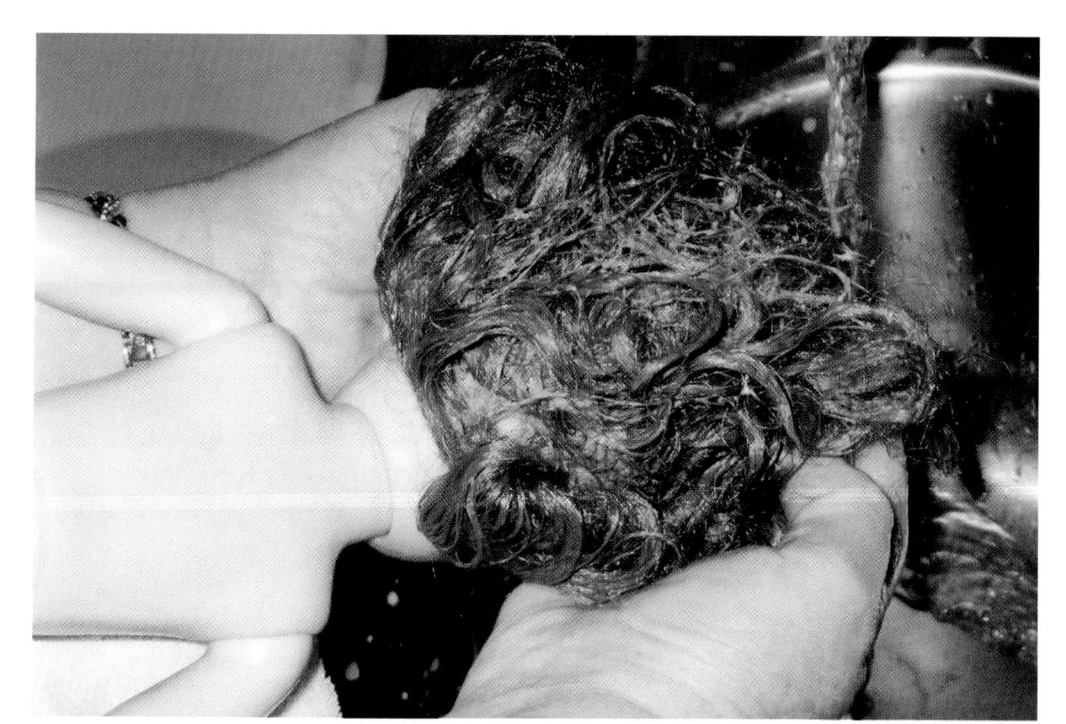

STEP #2B. Shampooing the doll's rooted hair.

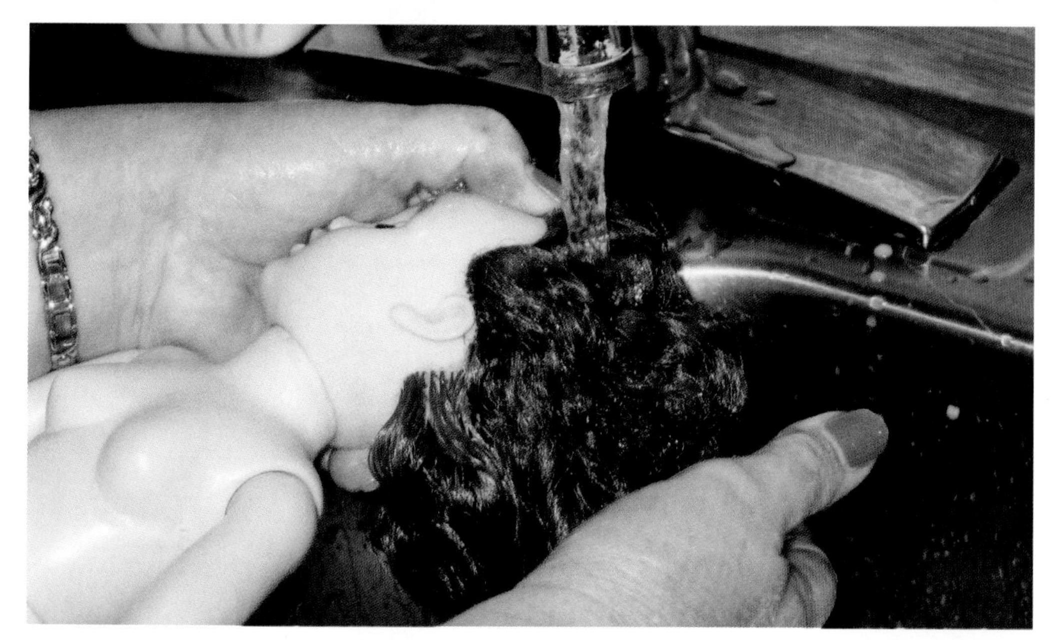

STEP #2C. Rinsing the doll's rooted hair.

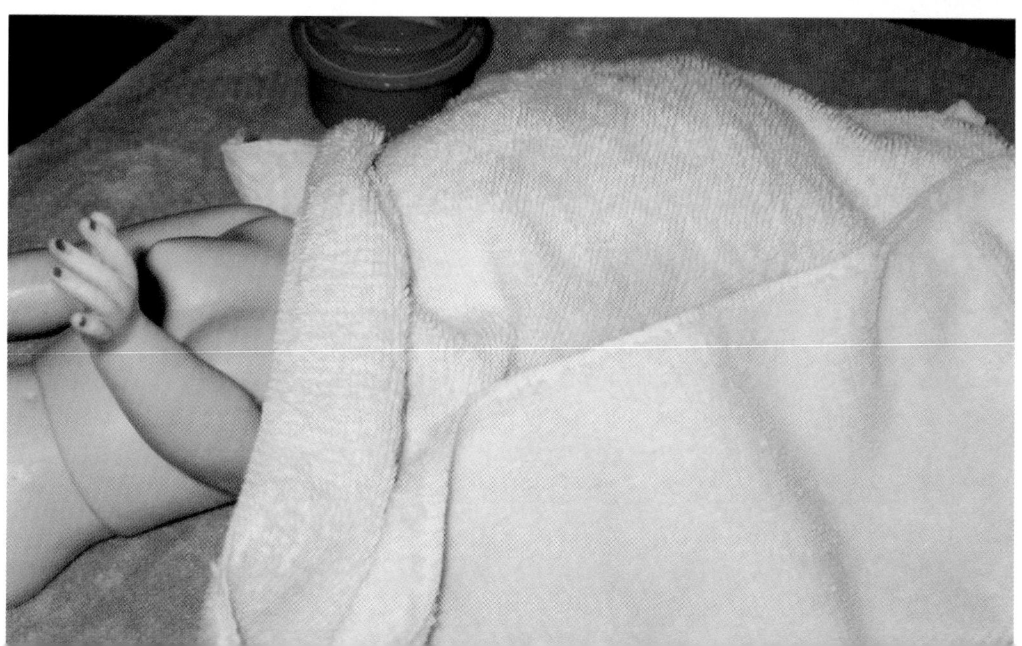

STEP #2D. Towel dry her hair.

Doll Wigs: Comparing Old and New

The inches displayed in the descriptions of the wig means the size of the head. Measure around the head above the eyes. (SEE A BELOW.)

Vintage wigs usually have a cloth cap inside them similar to old cheesecloth. Sometimes it is a natural beige color and sometimes it can be a shade of brown. (SEE B ON NEXT PAGE.) They were not made with elastic around the edge like wigs of today. (SEE C ON NEXT PAGE.) If you are seeking a vintage wig, shop for one close to the size of your doll's head. You can cheat a little with vintage wigs by clipping around the edge inward with the scissors. (SEE D ON PAGE **50**.) These slits will provide a little more room for a larger head or you can overlap the slits and sew them together for a smaller head. (SEE E ON PAGE **50**.) When the wig is glued on, it will not show.

Since the new wigs are made with an elastic cap as well as elastic around the edge, it is much easier to find a wig that fits the doll's head properly. When you see a wig marked "8-9½in" that indicates that it will fit a head that measures between 8in (20cm) and 9½in (24cm) around.

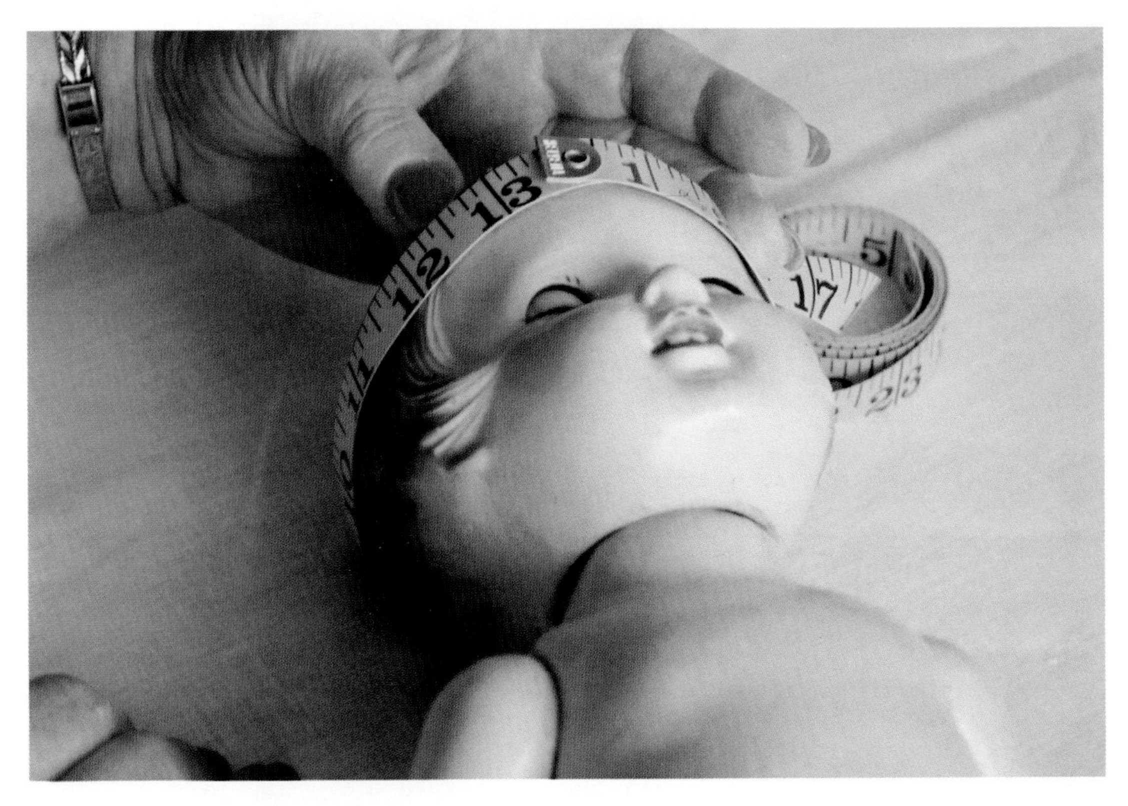

A. Measure the head for a wig.

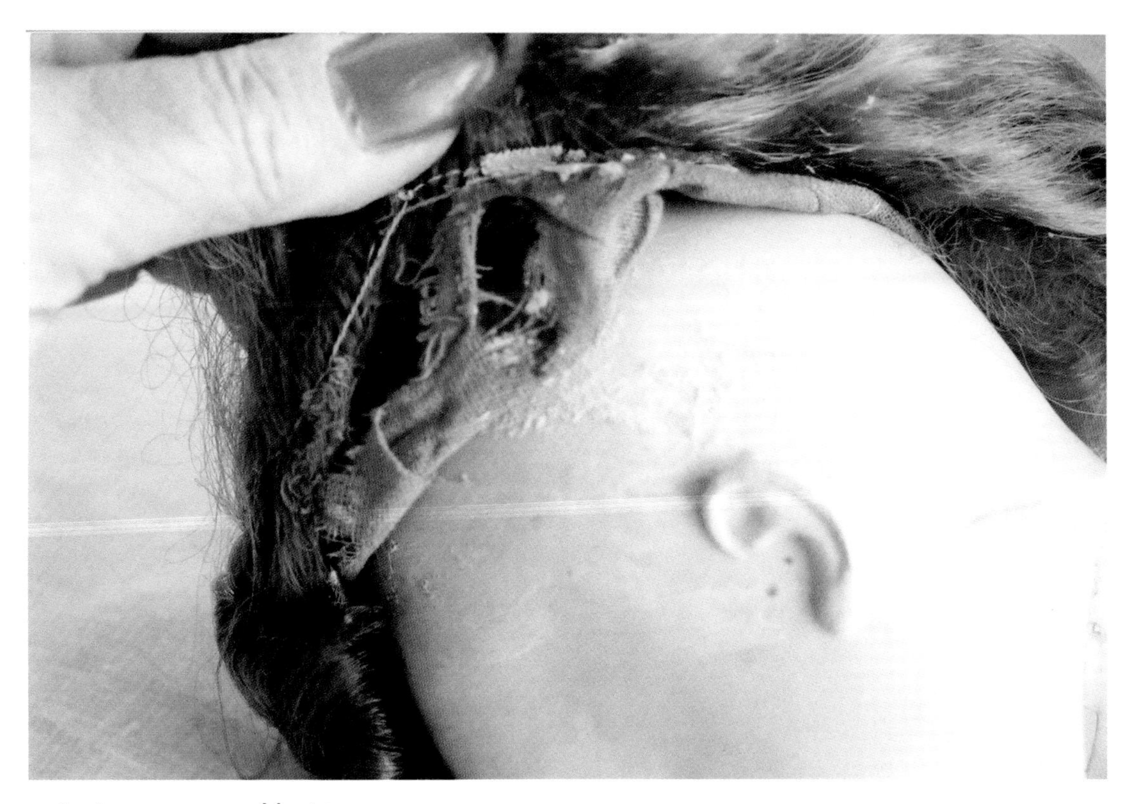

B. A wig cap on an **old** wig.

C. A wig cap on a **new** wig.

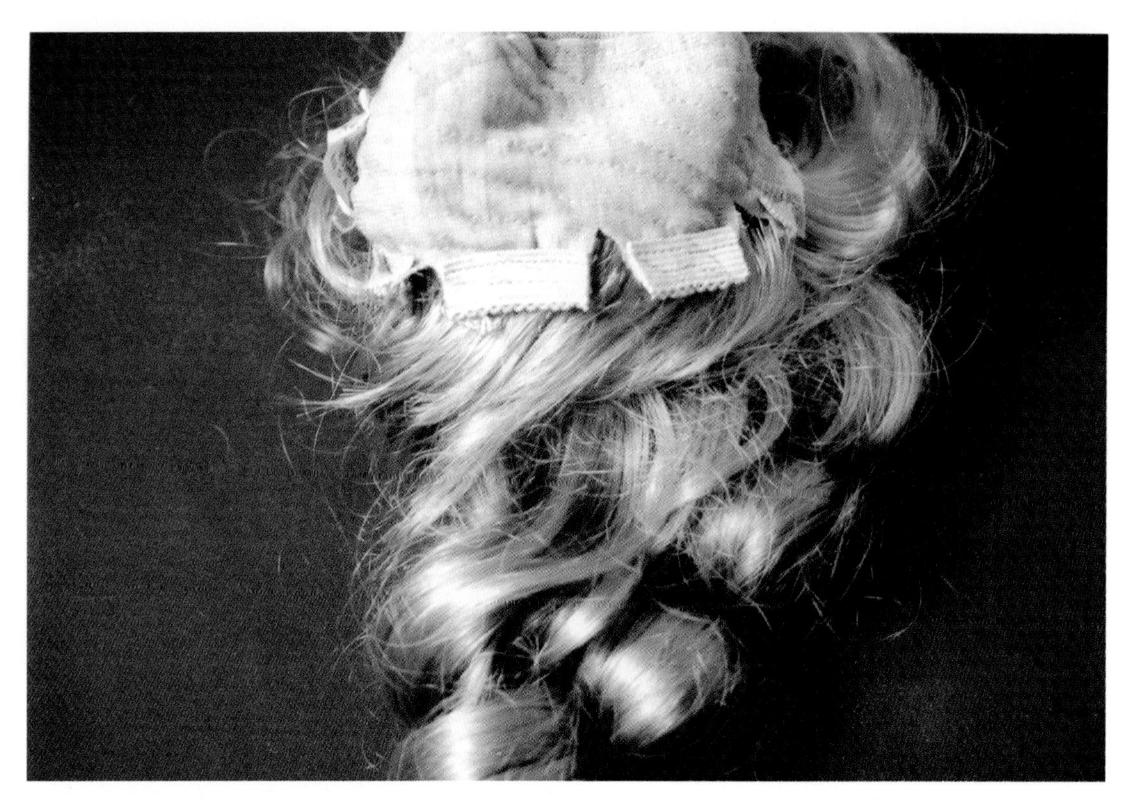

D. Clipping slits in the wig edge to enlarge the wig.

E. Clipping and lapping slits to make the wig smaller.

Removing the Old Wig and Glue on the Head

Before a new wig can be applied to your doll's scalp, many times an old one has to be removed. The glue that held the old wig on is the challenge. On occasion, I have resorted to using pliers and almost rolling them up with the wig clasped in them to remove the wig from the doll's head. This is done only when a wig is so bad that it cannot be saved for any future use.

The doll has probably gone through a lot of temperature changes in her lifetime, and this increases the probability that the glue has melted and possibly run down over her face and neck. The cleaner should be used to soften the glue. (SEE A.)

Once the glue has been softened, you will no doubt need to scrape or rub the area to remove it entirely. In the illustration, you will note that a very small area was wet with the cleaner on a cotton swab. The glue was easily removed from the area with a washcloth and a good rubbing. (SEE B.)

While the doll's head shown in the illustration is hard plastic, you can achieve the same results on bisque, composition or vinyl.

A. Wetting dried glue with doll cleaner.

B. The glue is removed in that spot.

Measuring a Doll's Head for a Wig and Applying It

1 To measure your doll's head for a wig, it is best to use a tape measure so you can hold it close. Measure around the head just above the eyes. (SEE **#1.**) You will note that all of the wigs usually span about 1 to$1\frac{1}{2}$in (3 to 4cm) so that the elastic band inside the edge of the wig will actually make the wig fit properly.

2 To put a wig on a doll, be sure that all of the old wig or any hair is removed from her head. If she had rooted hair which you are replacing with a wig, cut the hair off with scissors as close to the scalp as possible. (SEE **#2.**) This will allow the glue to adhere to the scalp.

3 Stand the doll between your legs with her facing your dominant hand. Open the wig up until it is inside out, holding it over your other hand.

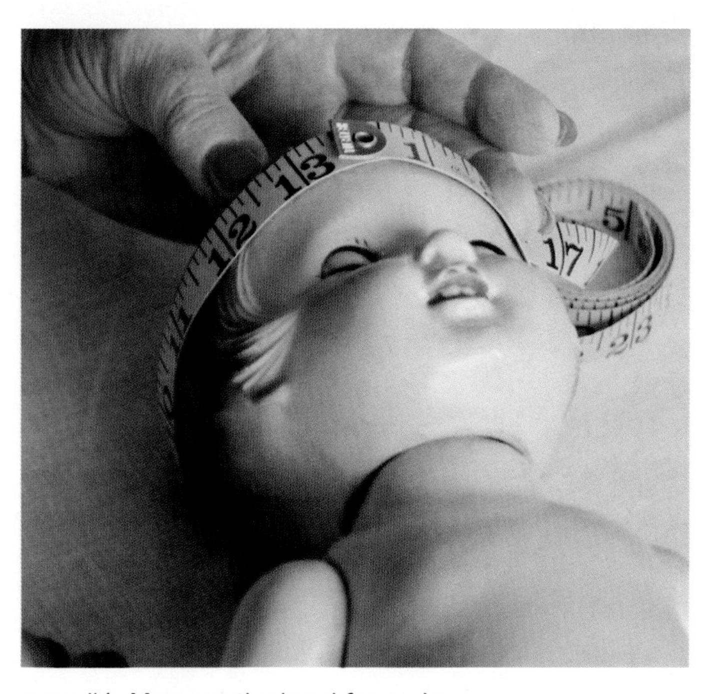

STEP #1. Measure the head for a wig.

STEP #2. Cut all old hair off.

4 Apply the glue (Aleene's Tacky Designer Glue works well, but Elmer's or any water-soluble glue also works.)

5 Apply dime size dots to the **inside** of the wig in five or six places, making sure that some of them are around the edge and some are in the center. (**SEE #5.**) The glue will not only hold the wig in place around her hairline, but will also give the wig a sturdy tight fit on top of the head. There is no need to spread the glue around, as the dots will hold it fine.

6 Hold the wig in your left hand (it is still turned wrong-side-out). Place the crown (very top middle) of the wig on the crown of the doll's head. Then begin to roll the wig down over the scalp. (**SEE #6.**)

STEP #5.
Applying glue to the underside of the wig.

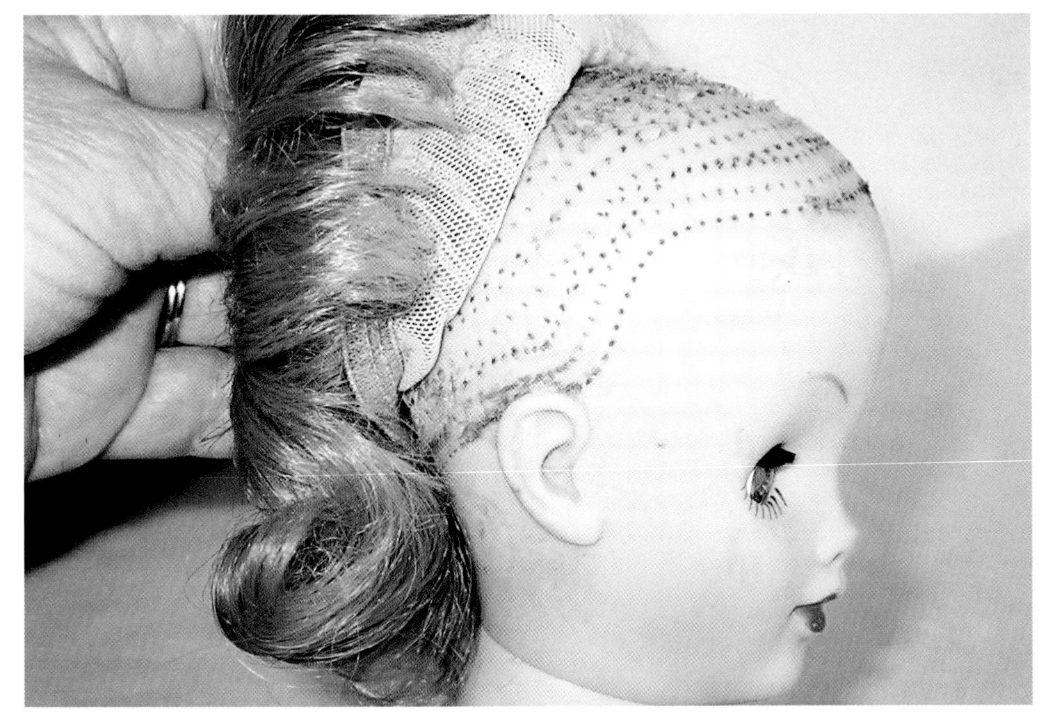

STEP #6.
Putting the wig on the doll.

7 The wig is now right-side-out on the doll's head. (SEE #7.) If you feel you need to adjust it (possibly back off the face a little) just take a pinch of the wig (and hair) and pull it around on the doll's head to straighten or make it look correct. There is no need to take it off completely and start over, as the glue will let it slide some. If you take it off and start over, you will likely get some of the glue on yourself or on the doll.

8 Check all around the doll's head and the edge of the wig by lifting up the hair to see that the elastic is turned right-side-down and in the correct position. It should cover where her hair was.

9 Use a wet cloth to wipe off any excess glue.

10 If a child is to play with her, let the glue dry about twenty-four hours first.

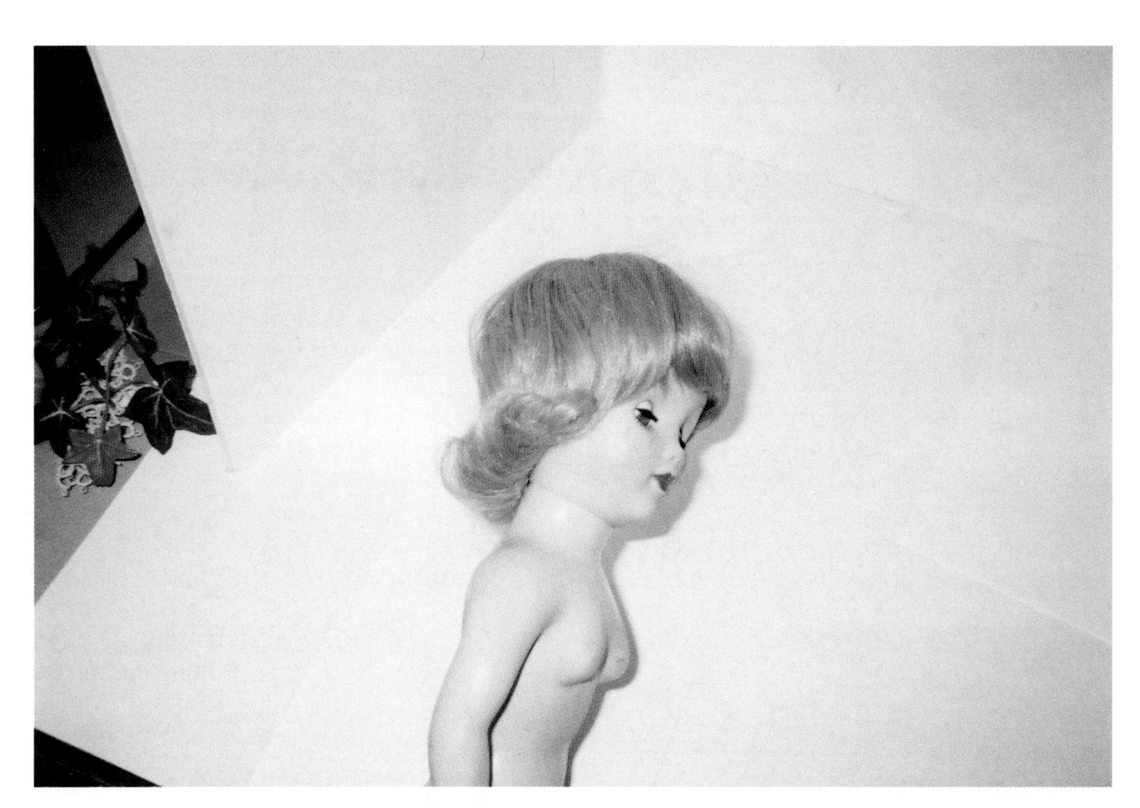

STEP #7. All done!

Restoring a Mohair Wig

When you have a mohair wig to restore, you will need plenty of time! Mohair wigs, while beautiful, are very fragile. Most antique bisque head dolls originally had mohair wigs or real human hair wigs. If the doll has been on display for any time or has been played with, chances are the mohair will be dirty and matted. Mohair can be made to look good again, but there must be enough of it left with which to work. (SEE PHOTO ON THE LEFT BELOW.)

1. Sit down with a towel over your lap and place the doll with the matted wig between your legs.

2. Using a toothpick, begin to "pick apart" the wig, strand by strand. (SEE **#2.**)

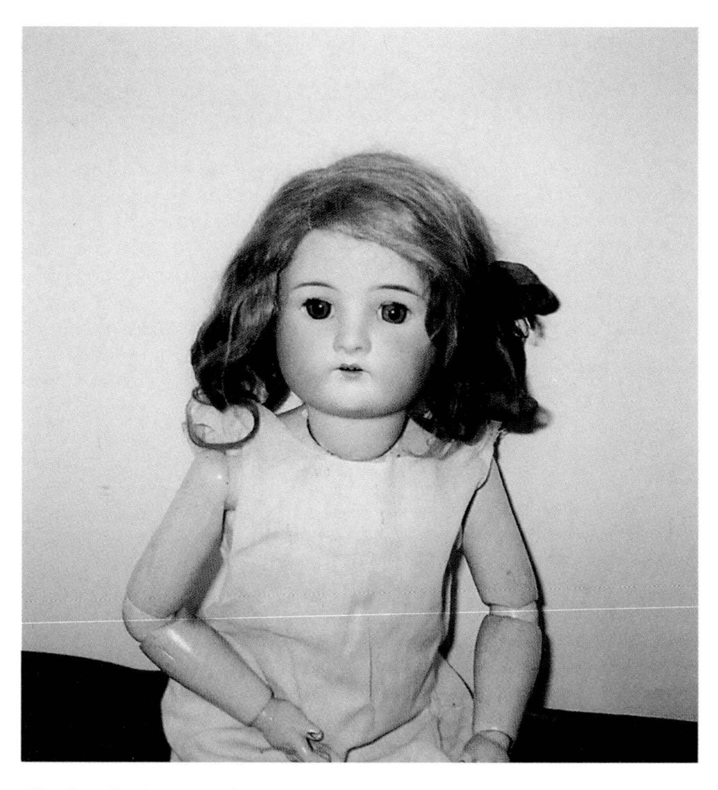

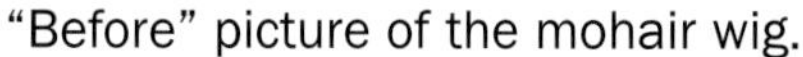

"Before" picture of the mohair wig.

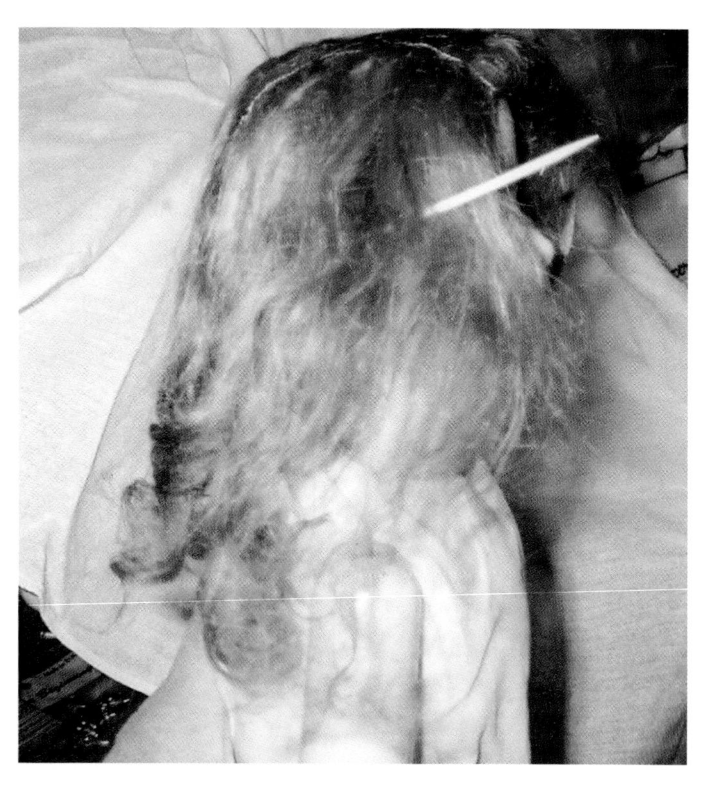

STEP #2. "Picking out" the matted mohair wig.

3 Pick it all over until you almost have an Afro look. You will need lots of hair to restyle the wig. (SEE #3A & #3B.)

4 Very carefully hold the doll's head under tepid water, letting it run through the wig, wetting all the hair.

5 Using liquid fabric softener — generously — as shampoo, wash the doll's hair. Squeeze the fabric softener in and out of the wet wig very carefully.

6 Rinse the wig well under tepid running water.

7 Pat the wig dry with a towel.

8 Let the wig dry thoroughly.

9 If any part of the wig is no longer attached to the doll's scalp, glue it down. Be sure to let the glue dry thoroughly before you go any further. (Aleene's Tacky Glue is great for use on doll wigs.) If you are working with the wig off the doll, pin it to a head form to work with it.

10 After the hair is dry, you may have to pick it apart with the toothpick again as the washing and patting dry may have re-matted the hair to some degree.

11 If the wig is sparse, sometimes it is best to barely touch it with a brush or comb. If this is the situation, smooth the top layer over on the doll's head. Leave some hair around her face and decide what style of bonnet or hat will look best on her to cover up her hair — or lack thereof.

STEP #3A. Lots more hair with which to work.

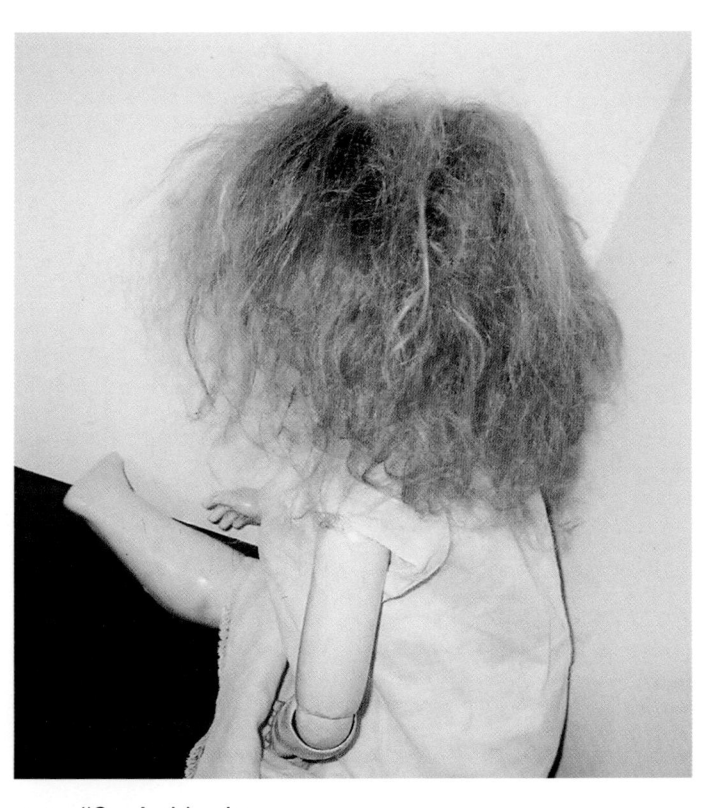

STEP #3B. A side view.

12 If there is plenty of hair, pick it out again to the Afro style.

13 You may carefully style it with a brush or comb but the best method of styling is to use home permanent curlers and papers to set it.

14 Part the hair into small sections for each curler, wetting the sections as you go with your fingers or comb dipped in water. (**SEE #14A & #14B.**)

15 Roll it up as if you were giving the doll a permanent. (**SEE #15.**)

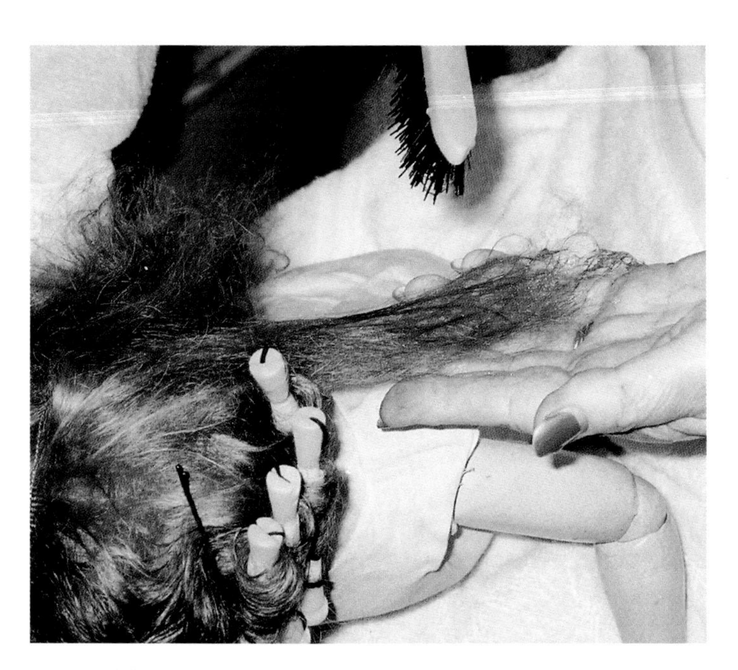

STEP #14A. Part off small sections of hair.

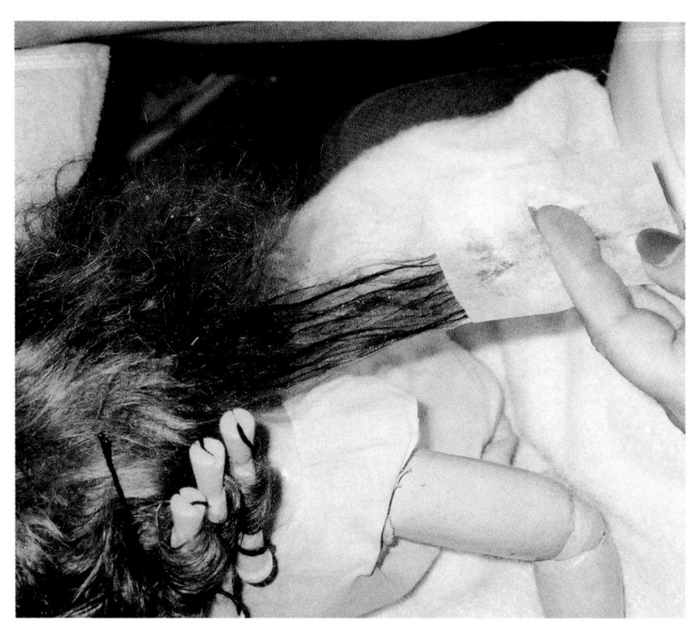

STEP #14B. Use home permanent curler papers.

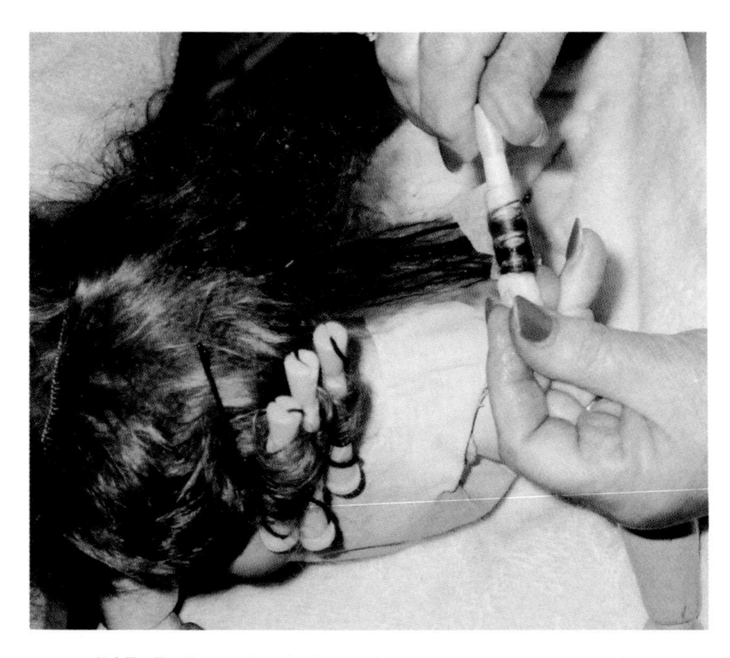

STEP #15. Roll up the hair on home permanent curlers using papers.

16 Decide if you want bangs or not. (SEE #16.)

17 Depending on the length of the wig, you may want long curls around her head or *Shirley Temple* style curls all over her head. (SEE #17.)

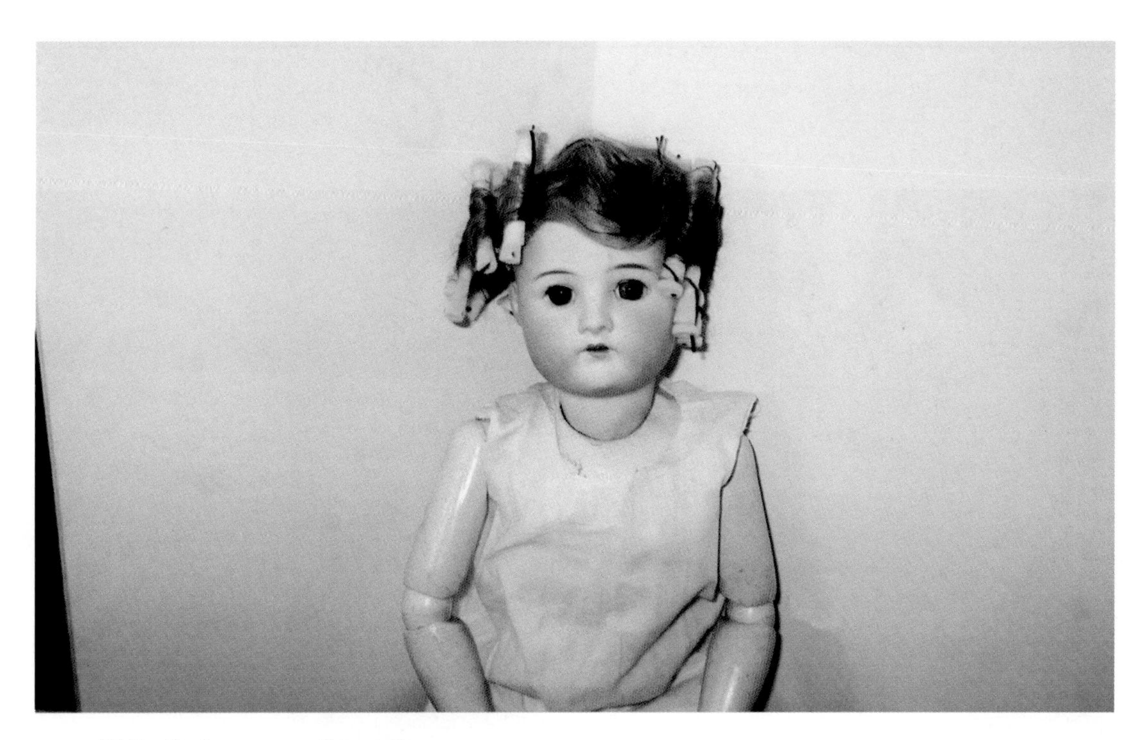

STEP #16. No bangs on this doll.

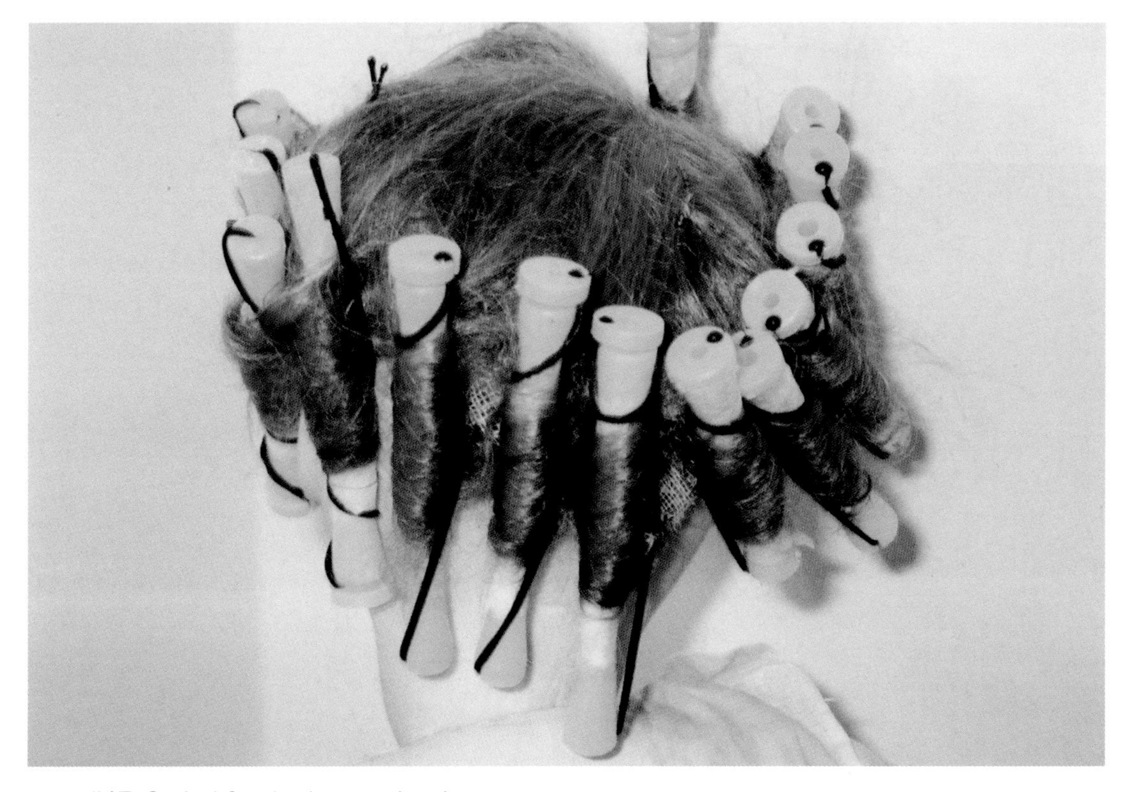

STEP #17. Curled for the long-curl style.

18 To make a style for an older lady doll, roll only the ends and bring them back in a cascade of curls down the back of her head. Leave the front and sides fairly smooth. (SEE #18.)

19 When the wig is thoroughly dry, very carefully unroll each curler and remove the end paper. (SEE #19.) Guide the curl back into place or just feather it out with your fingers into the desired style. Any **good** hairspray may be used at this time. (MINK™ brand is recommended, as it seems to make the mohair shiny again, in addition to holding it in place.)

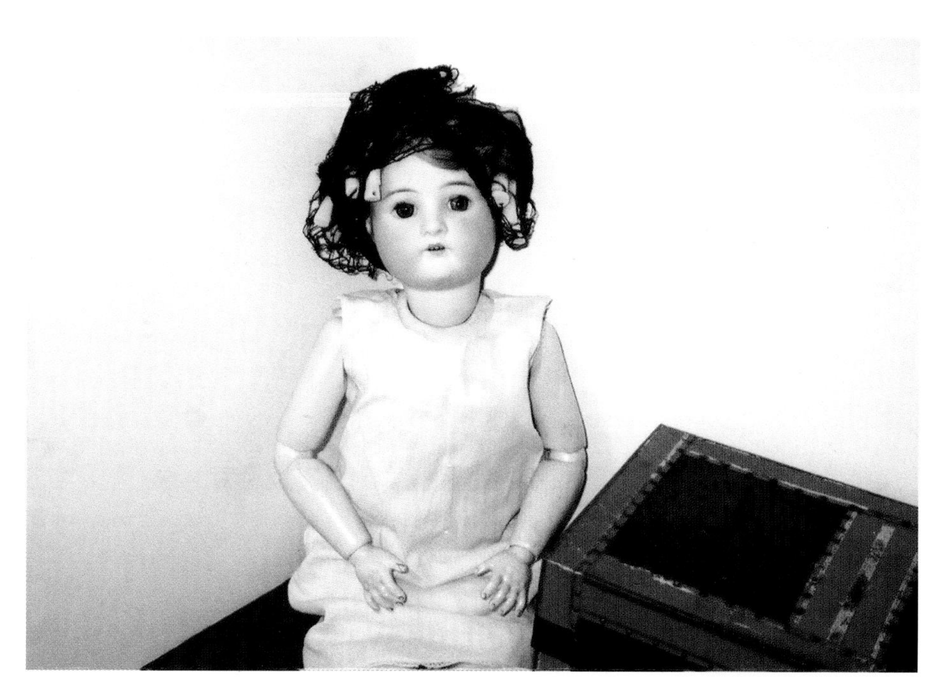

STEP #18. This girl is wearing an **old** silky hairnet to dry.

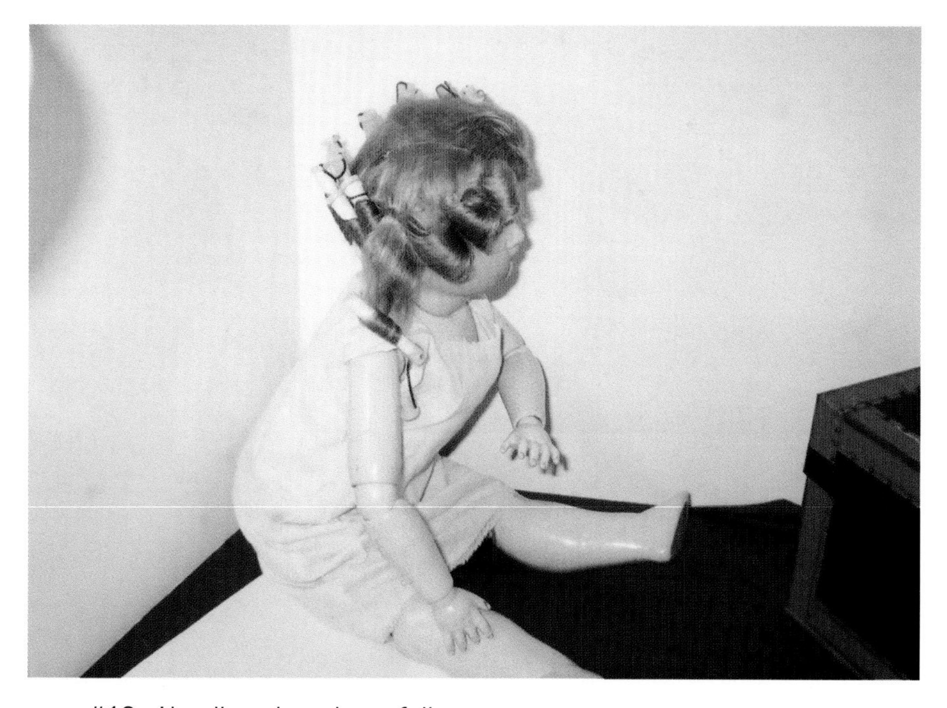

STEP #19. Unroll each curl carefully.

Now the doll looks beautiful once again.

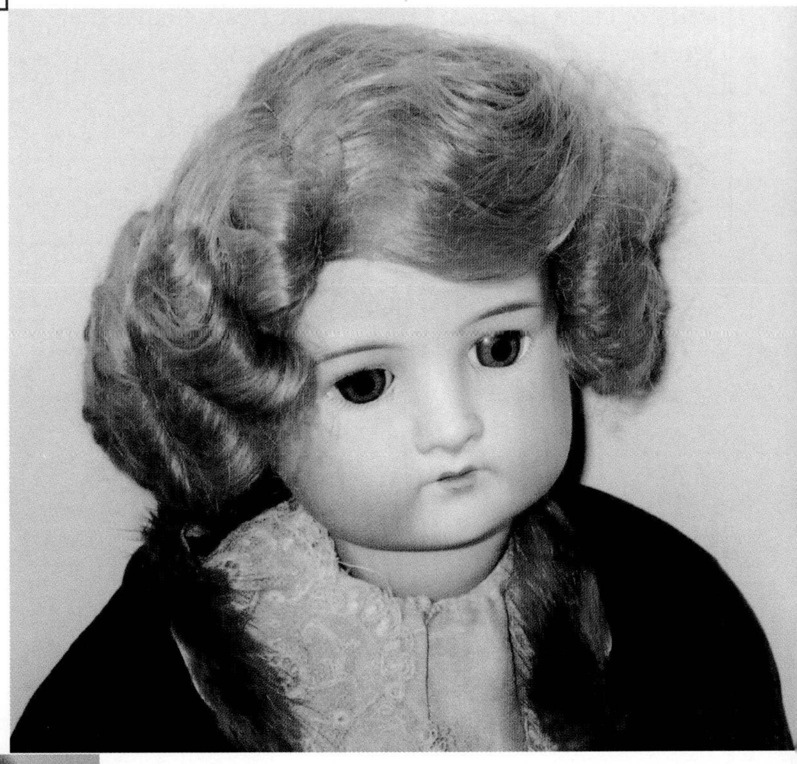

RIGHT.
The doll looks lovely!

BELOW.
A back view.

BELOW RIGHT.
Restoration completed!

Doll Bodies and Their Repair

Kid Leather or Faux Leather Bodies

Kid leather or faux leather bodies were used for early doll bodies. Many doll heads were made of bisque or china and were then placed on these bodies. (SEE PHOTO A, BELOW.) Sometimes the limbs were partially made of bisque or china and then were attached to the bodies with the leather. Some of these bodies were very crude but others were very well made with "gussets" at the joints of the elbows, knees (SEE PHOTO B, BELOW) and hips (SEE PHOTO C, ON NEXT PAGE).

Dolls with gussets were more poseable and could sit. Many times the feet and lower legs were made of cloth and were very stubby. These early bodies were generally tightly stuffed with sawdust. Over the years, the thread at the seams of these bodies gave way before the leather or cloth. They then lost sawdust readily because they were stuffed so firmly! Examine the illustrations again and you will see that the doll's problems are all at the seams of her kid leather body where the thread has rotted before the leather gave way.

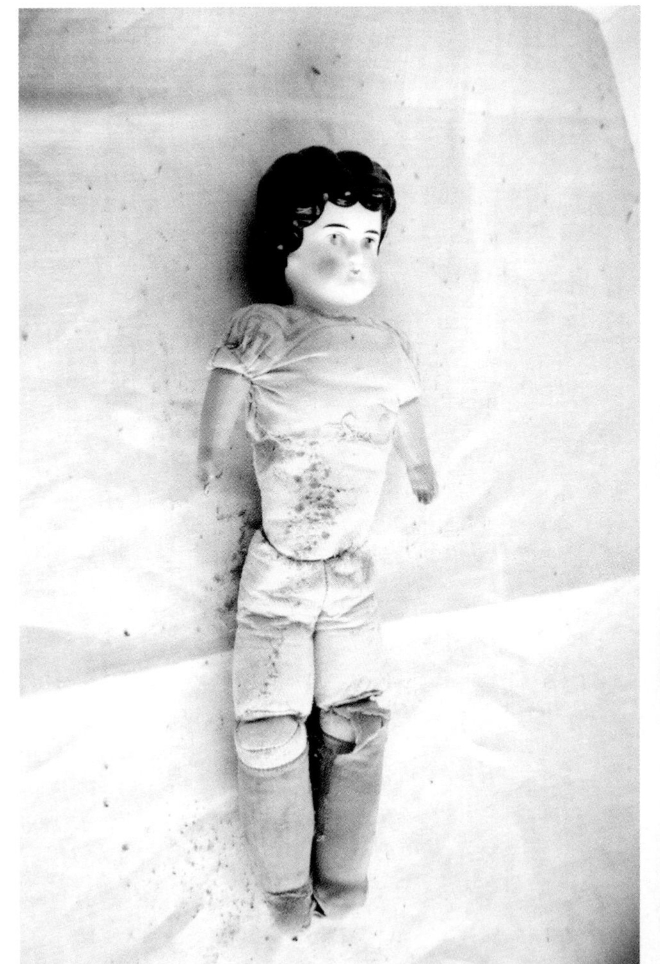

LEFT: A. China head doll on a kid leather body.

BELOW: B. Gusset in knee joint.

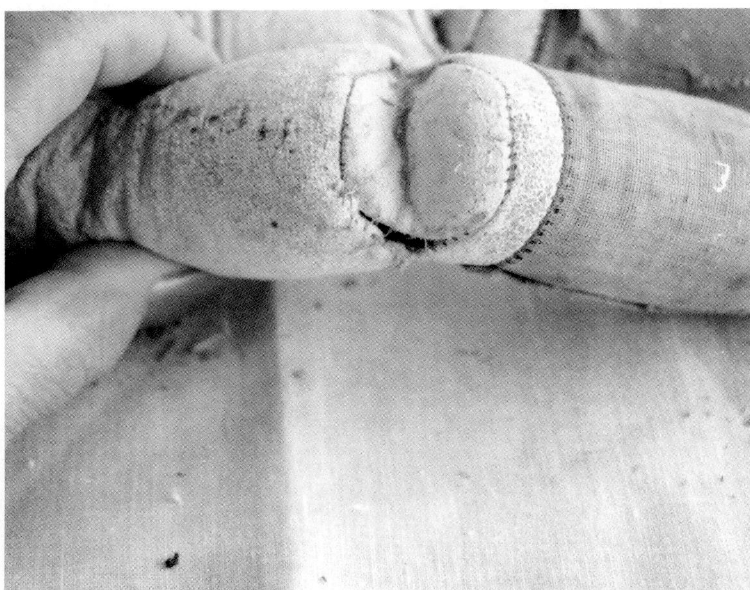

1 Kid leather gloves are needed for repairing these bodies. (SEE **#1**.)

C. Gusset in hip joint.

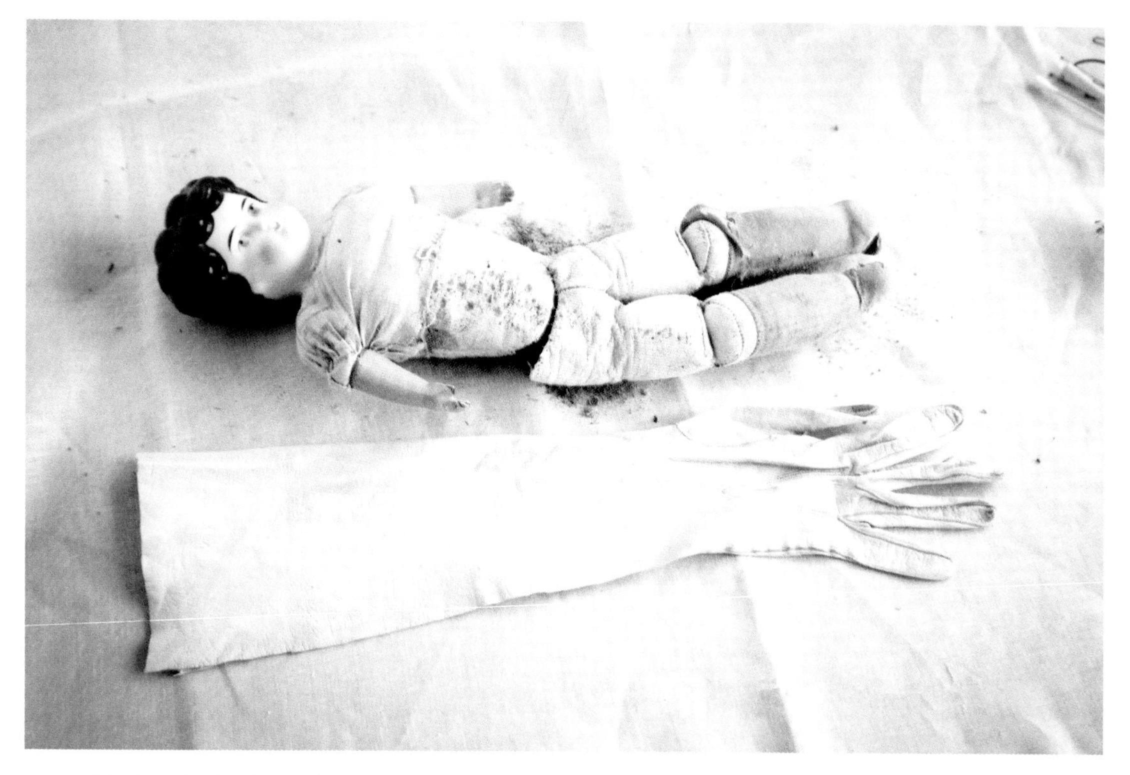

STEP #1. An old leather glove.

2 Cut a piece out of a glove the size and shape you need for the repair. (SEE **#2.**)

3 It can be glued on quite easily.

4 Round all the corners of your "patch" as you cut it. (SEE **#4.**) (The rounded areas will stay put better than a square corner as the corners will not get caught and pull up. Besides, most times you are working around a leg or arm anyway.)

5 Keep the sawdust from leaking by stuffing a cotton ball in the hole. Then glue on the patch.

6 The area has been strengthened and the patch you are gluing on will adhere nicely. (Aleene's Tacky Glue works best as it sticks very well for such a repair.)

7 All of the patches can be covered with her clothing – long socks and long sleeves cover a multitude of problems.

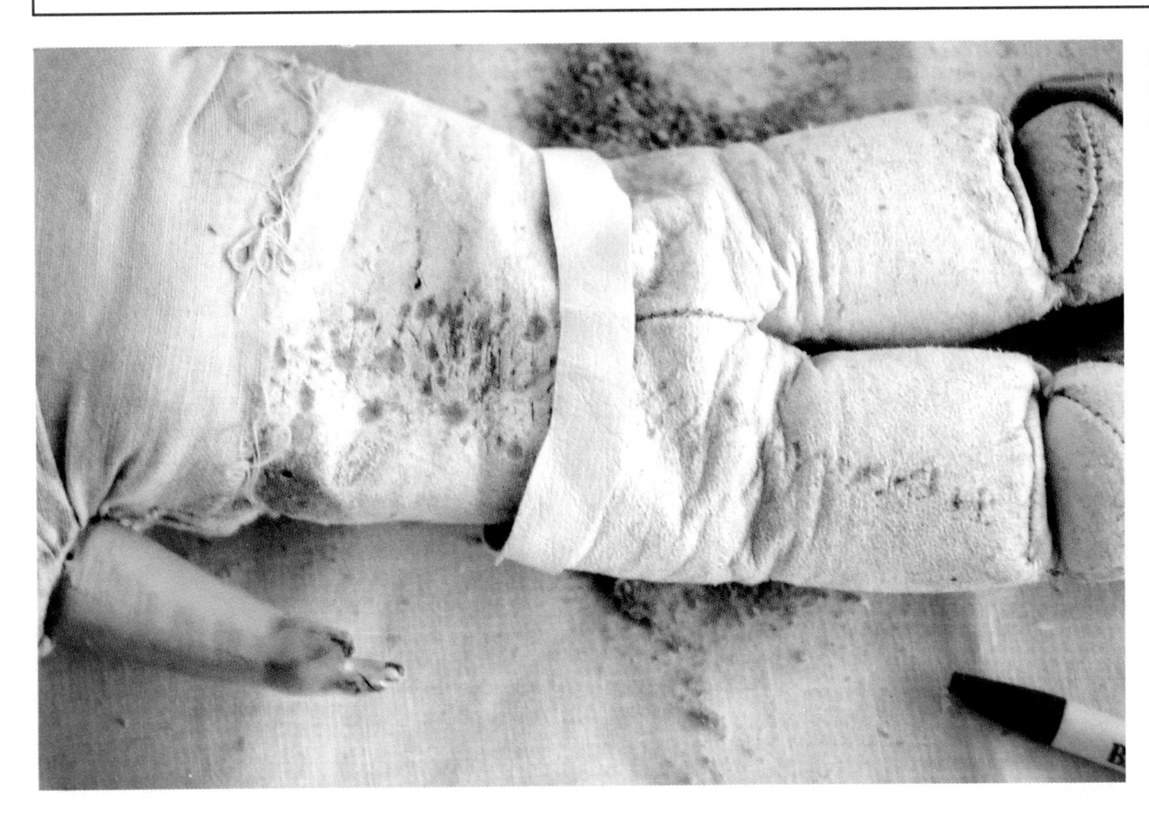

STEP #2.
Cut a strip from a leather glove

STEP #4.
A rounded patch from a leather glove.

Applying China Head and Limbs to a Cloth or Kid Body

Most china head dolls have porcelain or china lower arms and hands and some have china legs as well. However, many kid bodies are made with the legs and feet out of kid leather or cloth. Kid bodies usually have kid upper arms. Cloth bodies have cloth upper arms. (SEE PHOTO BELOW.) The upper part of the arm should extend from the shoulder to the elbow or forearm in order to accommodate the porcelain or china lower arm.

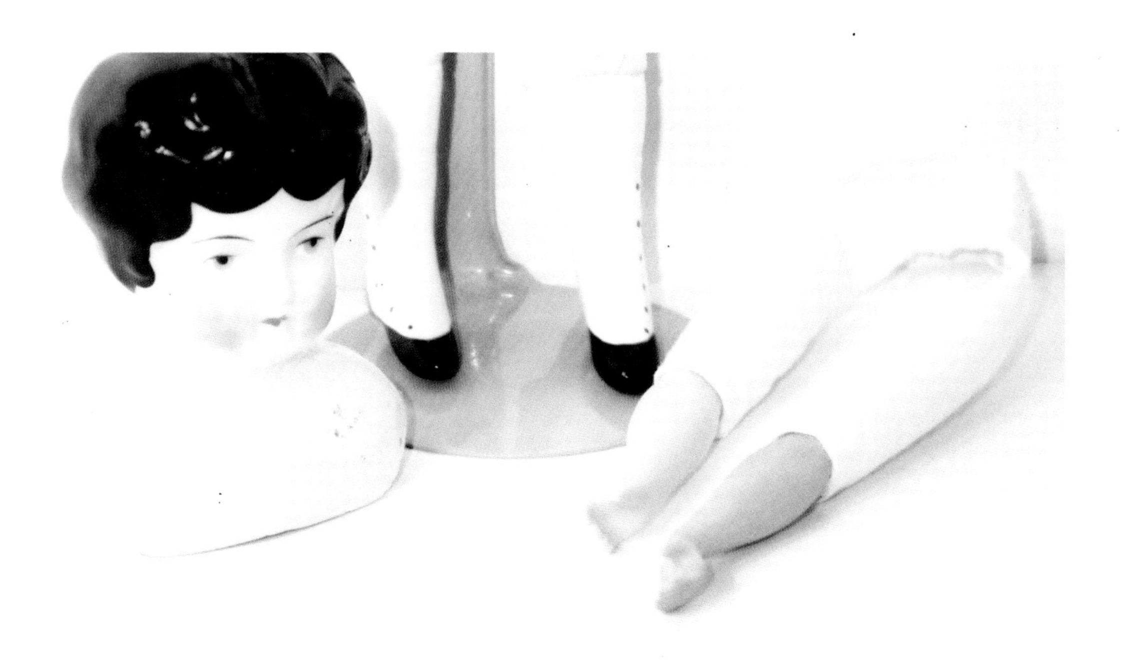

The doll arms are a good length.

1 The kid arms can be constructed from scraps of kid gloves. They each have a seam sewn at the underside of the arm. (SEE #1A & #1B.)

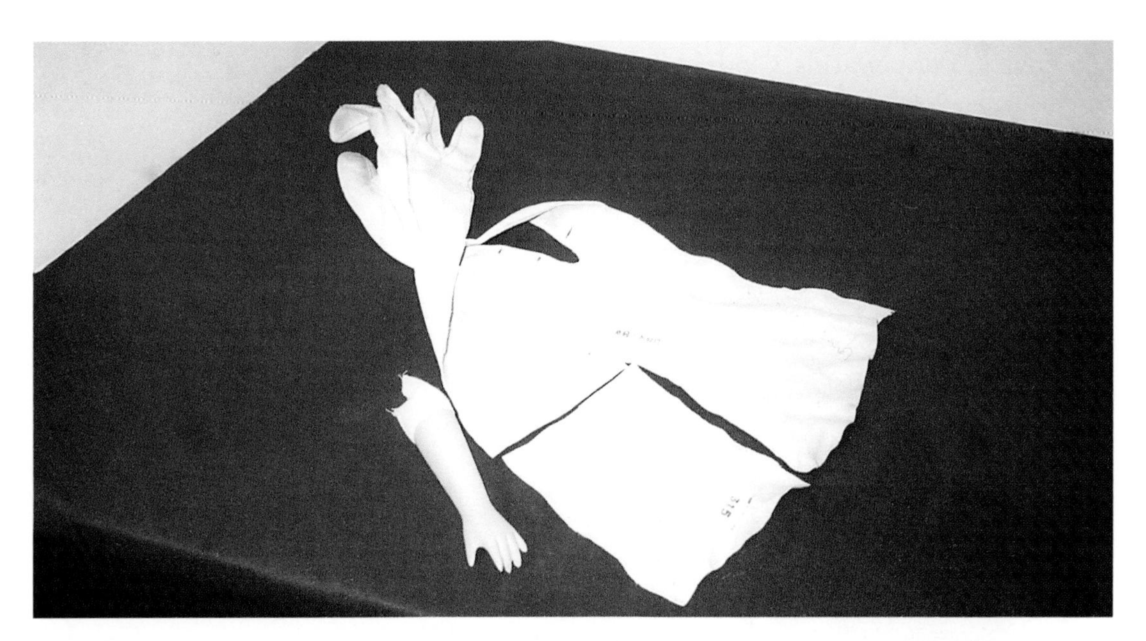

STEP #1A. Cut an upper arm piece from a leather glove.

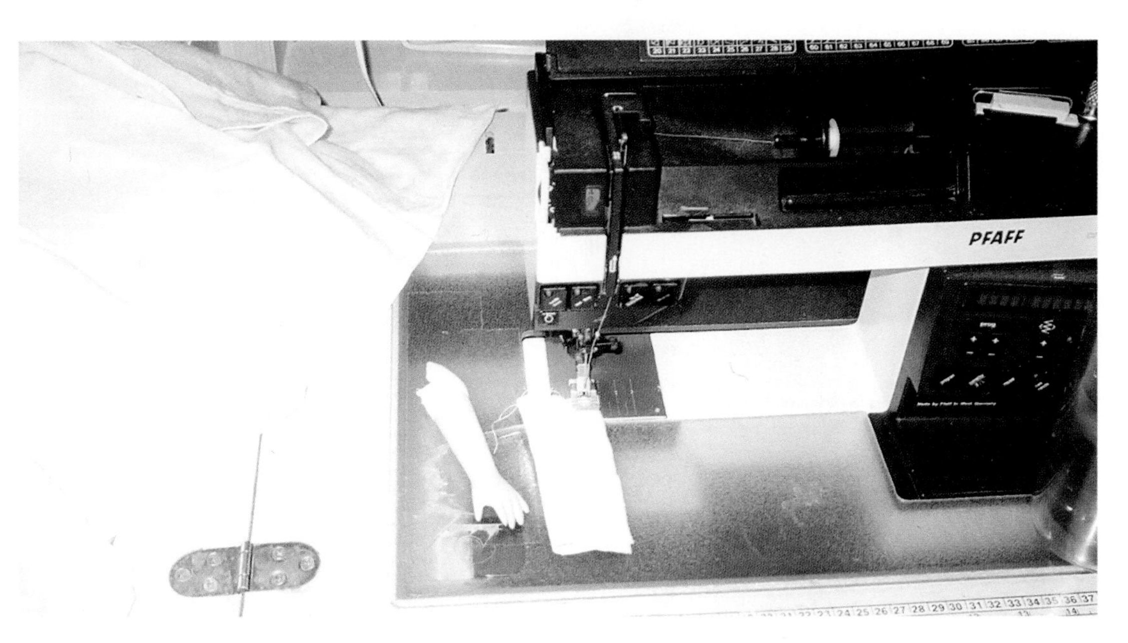

STEP #1B. Stitch the arm seam on the machine and turn it inside-out.

2 Apply glue to the lower arm at the elbow area so that you can insert it into the upper "sleeve" of kid leather.

3 Let it dry thoroughly. (SEE #3.)

4 Stuff the upper part of the kid leather arm with polyester fiber filling or cotton. (SEE #4.)

5 Use the same steps —steps 1 through 4 — when making cloth upper arms for a cloth body.

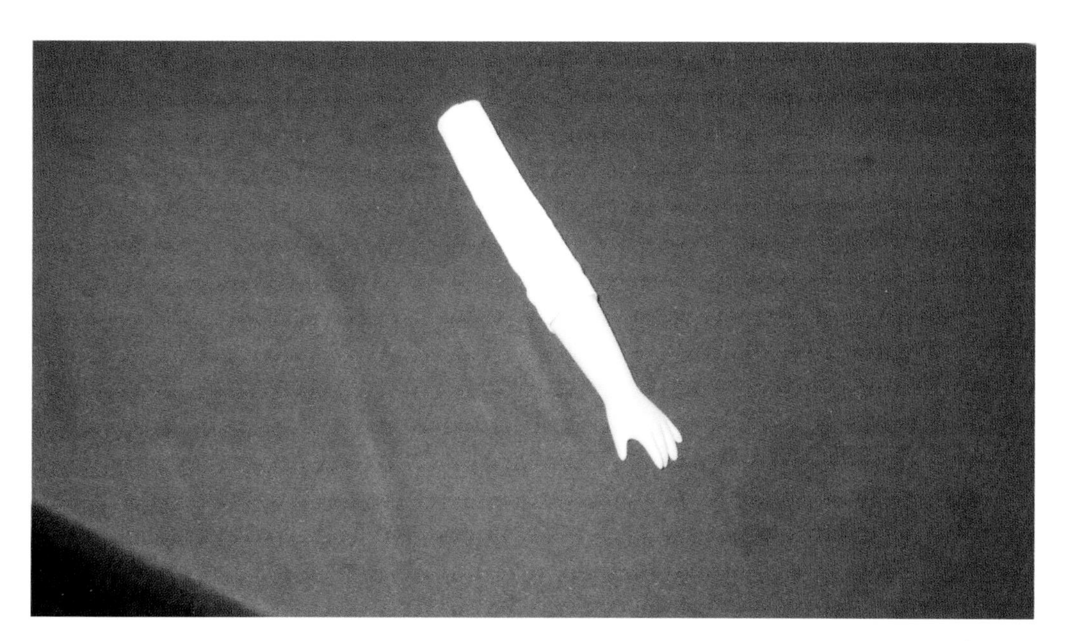

STEP #3. Put glue on the arm and insert it in the new leather upper arm at the elbow. Let it dry well.

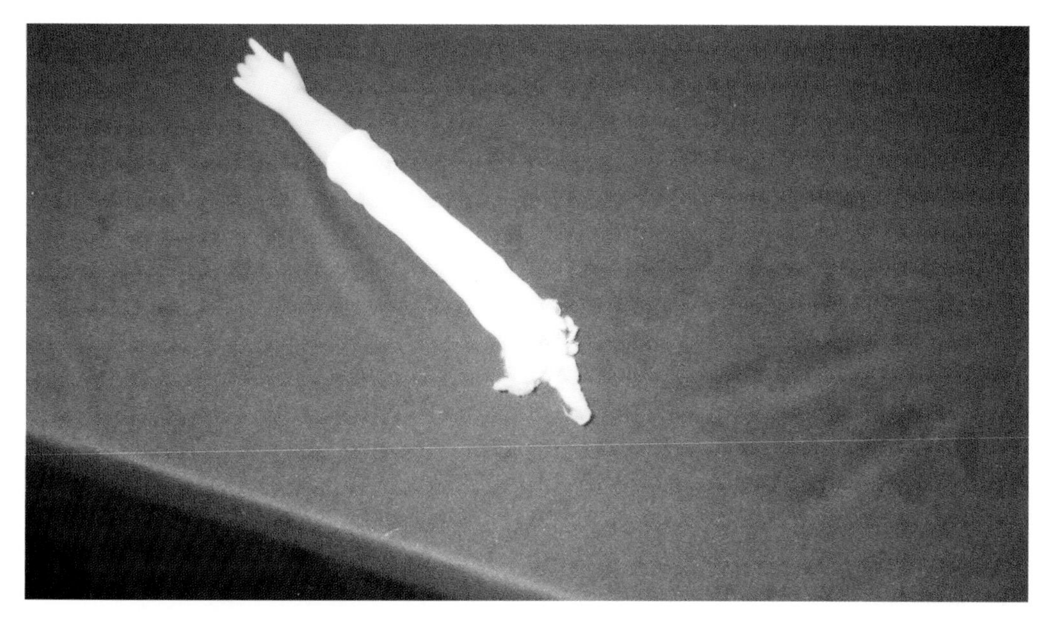

STEP #4. Stuff the upper arm.

6 Always place the upper end of the arms on the stump of the torso to gauge where and how long the completed arms should hang. The hands should be level with the hips or just a little above. (SEE **#6.**)

7 Measure the distance **between** the arms where they are placed on the top of the torso.

8 Cut a piece of cloth (any kind, as it will be hidden) to fit between the upper arms.

9 Sew the upper arms to it, thus joining them. (SEE **#9.**)

10 Place this joined piece **across** the torso before you apply the shoulder head. (SEE **#6** AGAIN.)

11 The fabric will hold the arms in place. (The old way of doing this was to sew the upper arms to the body by hand. Years later, the stitching rotted and the arm came loose. Many times arms were lost or broken from falling off. The dressing and undressing also contributed to the loosening of the stitches. By joining them together, as described here, you can prevent this from happening.)

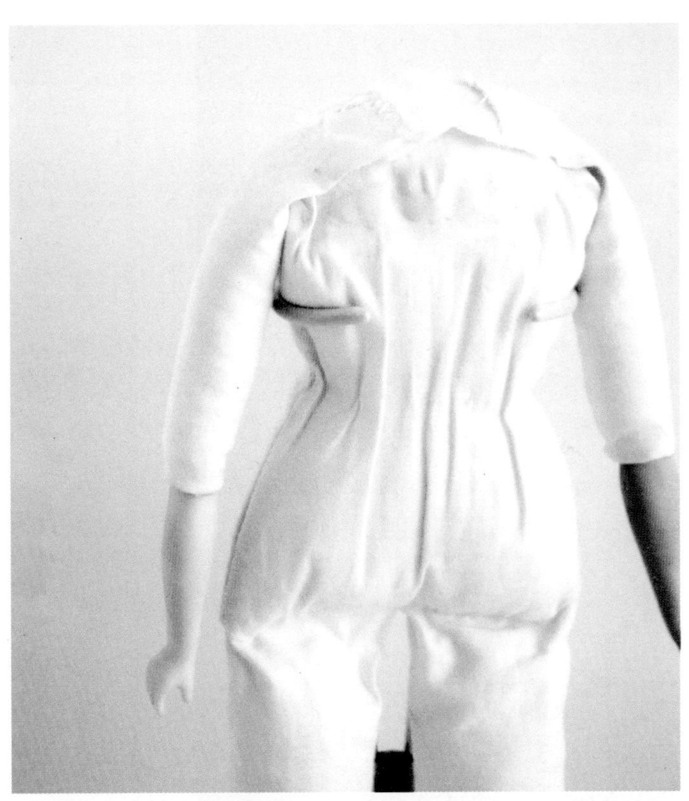

STEPS #6 & #10.
Place the arm on the body to determine the correct length of the arm.

STEP #9.
Stitch a piece of cloth to each upper arm, joining them.

12 Next, attach the shoulder head to the body.

13 Spread some glue on the inside of the shoulders near the bottom edge, on both the front and back. (A thin line of glue is sufficient.) (SEE **#13.**)

14 Push the shoulders down on the stuffing in the top of the body. (SEE **#14.**)

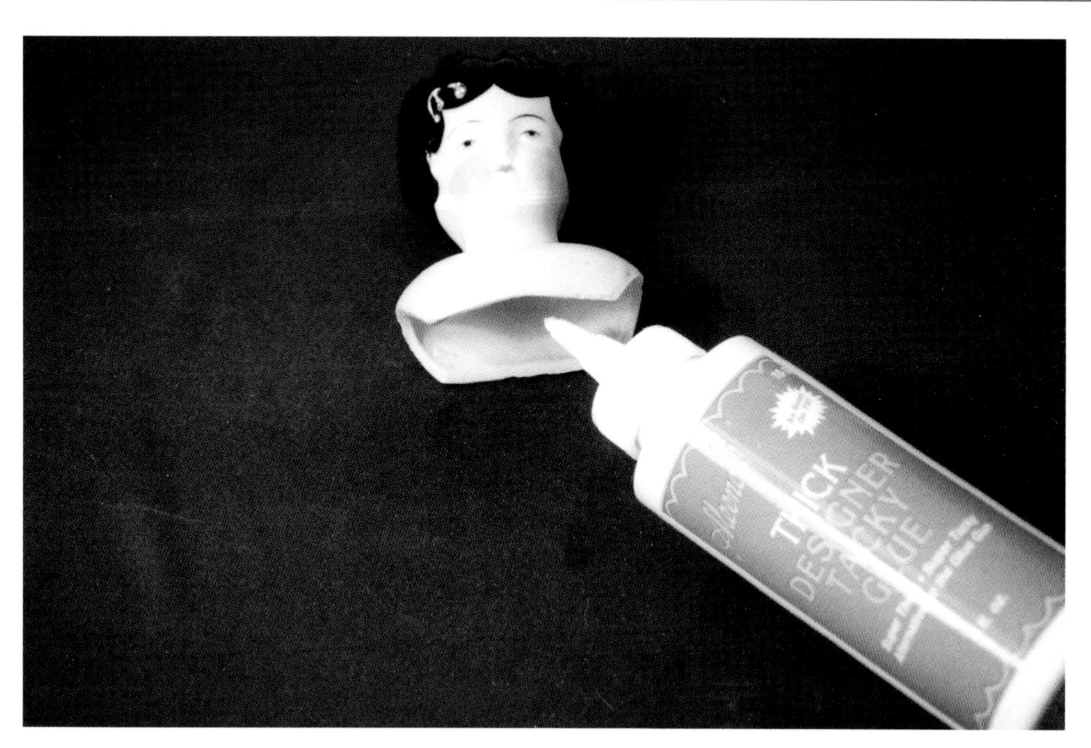

STEP #13. Apply glue to the underside of the shoulder plate, both the back and the front.

STEP #14. Place the head and the shoulder plate on the body.

15 Push and pull the body and the arms around the shoulder plate to make it look even and level on the body. (SEE #15.)

16 A kid body is made a little different at the top. It usually has straps that open at the top of each shoulder allowing you to insert the shoulder head down **inside** the kid body. The straps of kid leather will meet and overlap on the shoulders from the back and the front.

17 Hold the body between your legs.

18 Put the head in place.

19 After placing the shoulder-head into the proper place in a kid body, squeeze some glue out onto a scrap of paper. (Aleene's Tacky Glue is recommended, but any glue will do. Do not use a glue gun.)

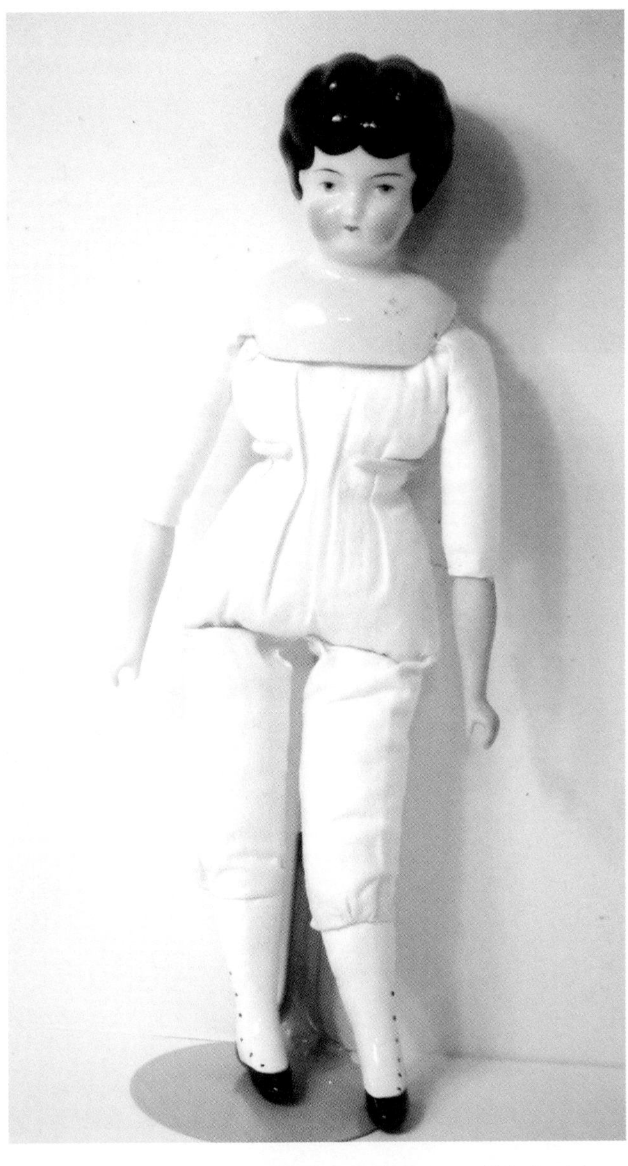

STEP #15. Check the body to make sure all is inside the shoulder plate in the front and the back.

20 Use a toothpick to smear it on the shoulders where the straps will lay and where they overlap.

21 Place the front strap on top of the back one when you finish. If they do not overlap, it is okay as sometimes the body does not have straps that are long enough.

22 If you feel the kid leather is not as supportive as you want to hold her safely, you can cut a long narrow strip of kid leather from a glove. Glue the strip to span the edge of the kid leather and the china shoulder plate.

23 In the illustration, you will see that a long strip of kid leather was used to completely cover a seam on the doll's stomach where it had separated. (SEE **#23.**)

NOTE Do not try to use a needle and thread to simply mend a rip in a kid leather body as you will find that the thread pulls right through the kid leather because it has deteriorated with age. You will be grateful to have old kid gloves on hand to cut up for patches.

STEP #23. Cut a strip from a leather glove.

A New Cloth Body

The following instructions are for replacing a cloth body on a doll with vinyl arms, legs and head. (SEE ILLUSTRATION ON THE RIGHT.)

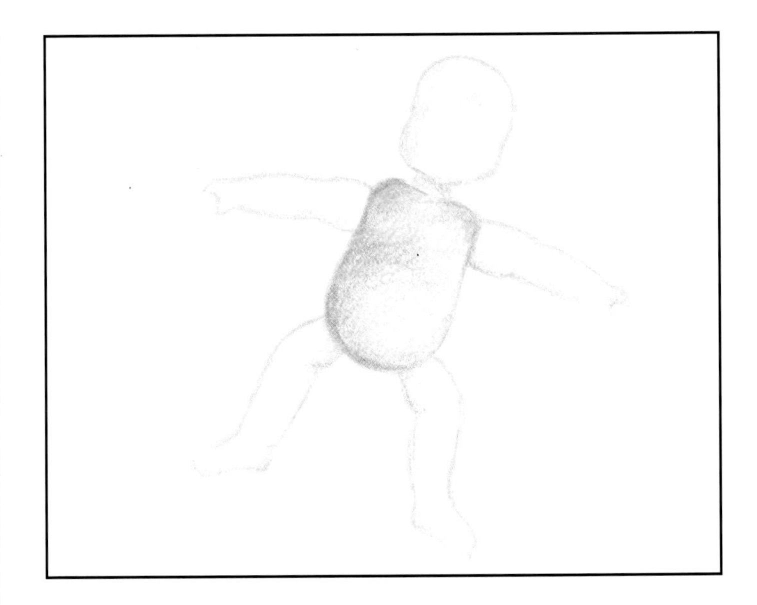

1. Dismantle the doll by carefully removing the cloth body from the limbs.

2. Take the body apart at the seams and steam press the individual pieces.

3. Using the old pieces for a pattern, cut out the new body. (Be sure to save those pieces of cloth body that you used for a pattern. You will soon have a collection of many sizes of body patterns.) (SEE **#3.**)

4. Try to use the same color (pink, off white, white or dark brown fabric) as the original body.

STEP #3. Use the old cloth body pieces for a pattern.

5 If this is a doll that a child is going to play with a lot, it is a good idea to use two thicknesses to be sure it will last.

6 Prepare to sew the body together. (SEE **#6.**)

7 Sew the shoulder seams and the side seams. (SEE **#7.**)

STEP #6. Sew the shoulder seams and side seams.

STEP #7. Clip the curves of any seams.

8 Sew the binding around the neck so you will have a casing to run either a cable tie or a wire through to hold the doll's head on. (SEE #8A & 8B.)

9 Leave the top of the back of the body open from the neck to below the waist.

10 Sew the rest of the back seam together.

11 Sew the crotch of the front and back together.

12 Since the old body was sewn to the arms and legs, try to find a shoe repairman who will stitch the arms and legs into the cloth body for you.

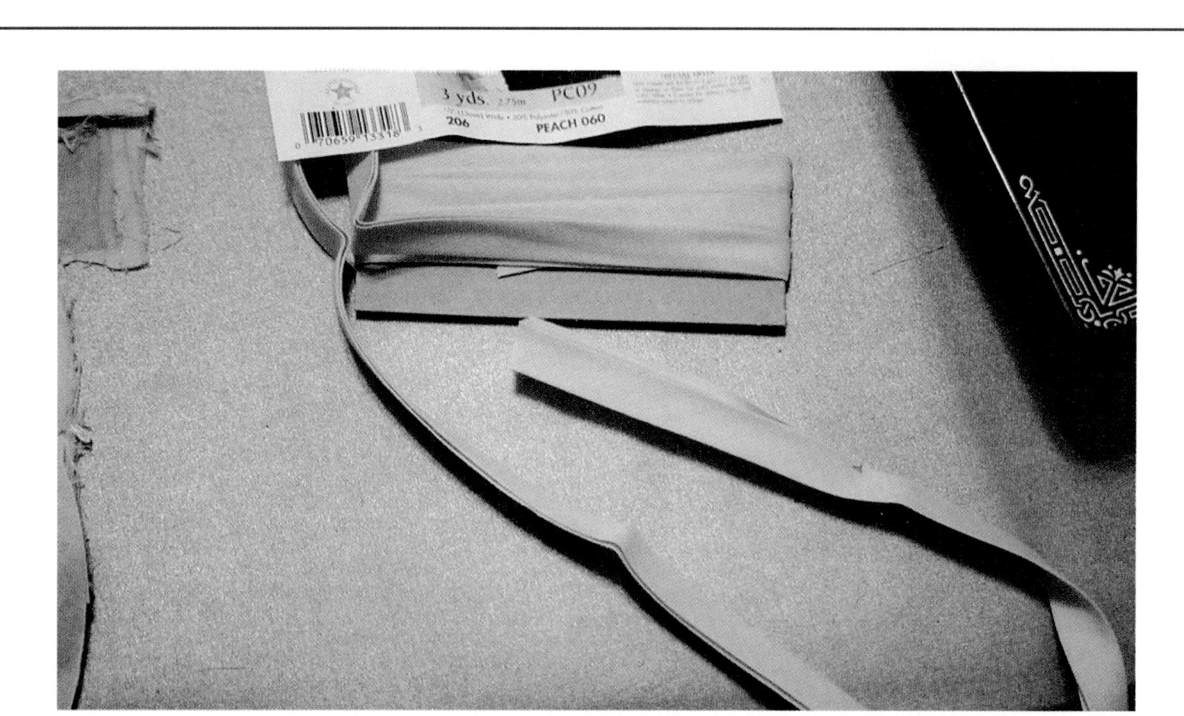

STEP #8A. Use bias binding for the neck.

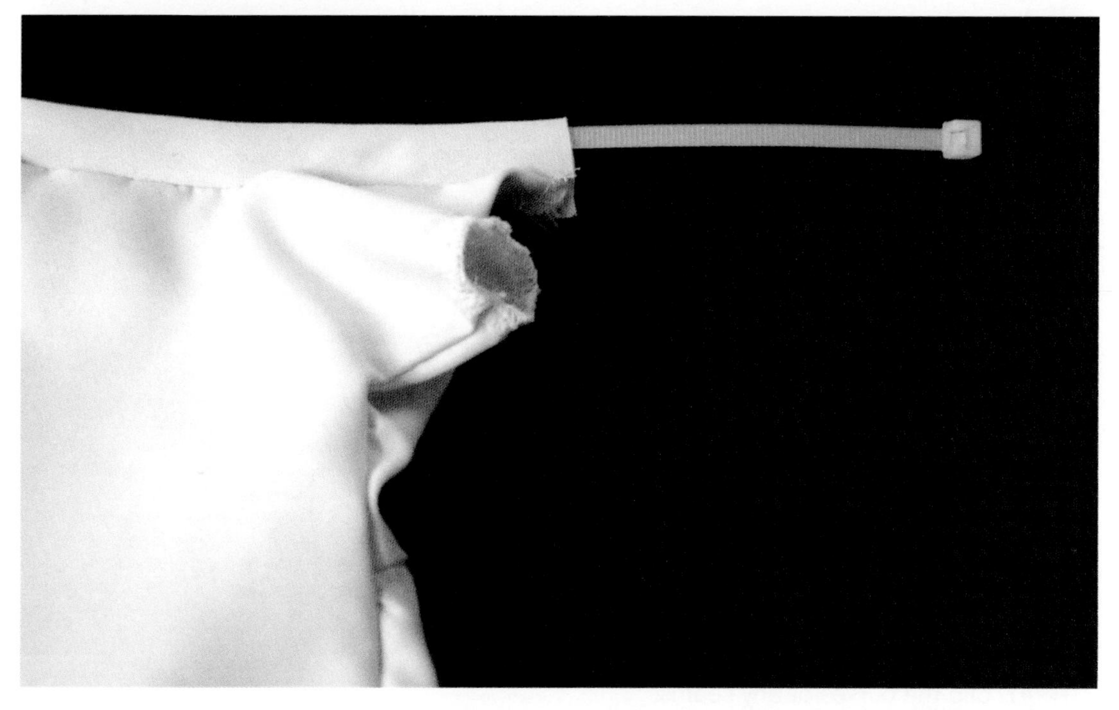

STEP #8B. Sew bias tape to the neck and insert a plastic cable tie in the neck casing.

Preparing the Cloth Body and Limbs for the Shoe Repairman to Stitch

1 Put the arms through the armholes.

2 Pin them to the cloth using straight pins. (SEE #2.)

3 Do the same with the legs.

4 The sewing machine that the shoe repairman uses will fit inside those little limbs. All he has to do is sew right around on the inside of the leg or arm and he is through but you must have it all ready for him, as this really is not his line of work. He just might not want to take the time to do this if you have not made it simple for him. (My shoe repairman charges about $2.00 a limb.)

5 **Very important –** the arms and legs need to go into the holes in the newly made body a certain way. Place them correctly so that when you turn the body right-side-out you do not have one arm trying to scratch the doll's back while the other rests in her lap! If that happens, it means it has to be done over again, including another visit to the shoe repairman.

6 Always use several straight pins when securing the limb to the body opening. (SEE **#2** AGAIN.) At least four should be used in order to have the limb secured evenly, but six or eight in each limb makes the shoe repairman's job easier! Pliers or a strong hemostat are necessary to hold the pins and push them through the vinyl and the cloth to stay. (SEE **#6**; LARGER PHOTO ON PAGE **76.**.)

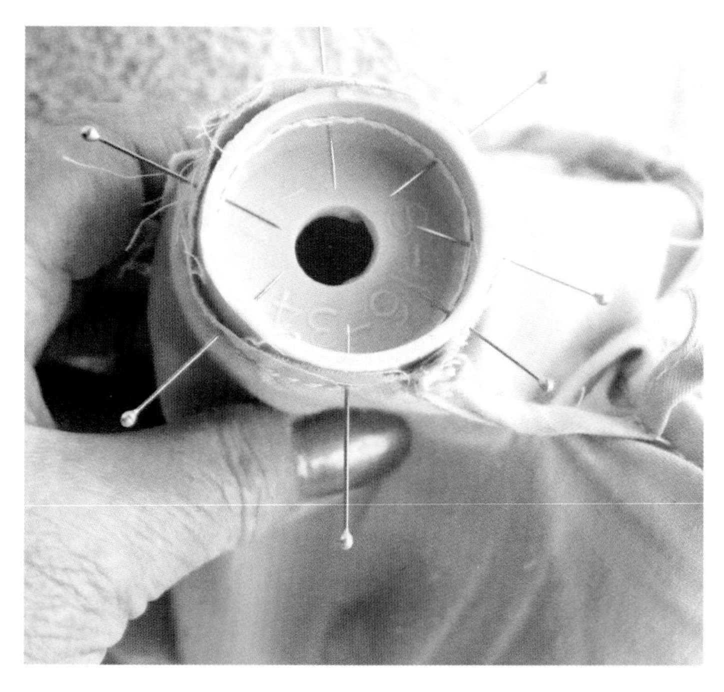

STEP #2. Pin the limbs into the body.

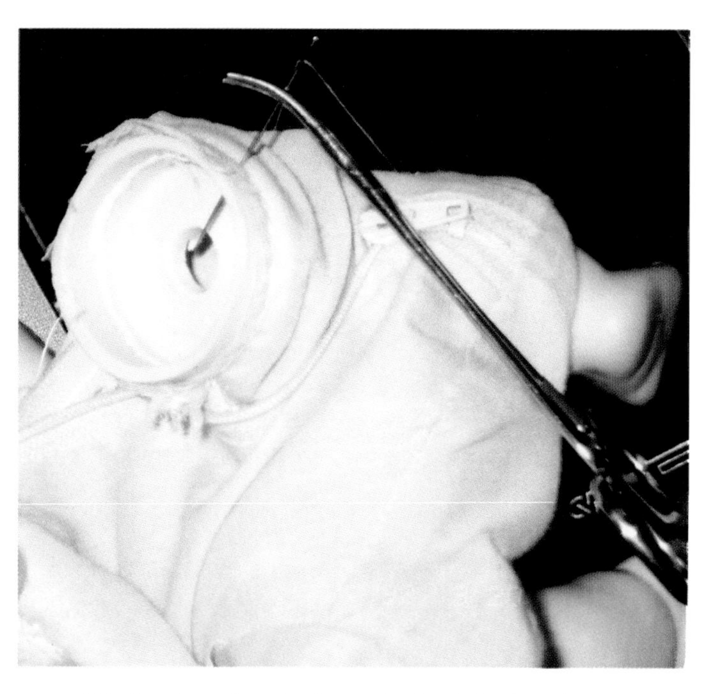

STEP #6. Sew the limb to the cloth body by hand.

Reattaching a Limb to a Cloth Body

1. If you ever have a very small rip that you need to restitch on a limb, use a strong needle and strong thread to fix it by hand.

2. Try to go back and forth in the holes already in the vinyl where it was originally stitched. (SEE **#2.**)

3. Do this several times in each hole.

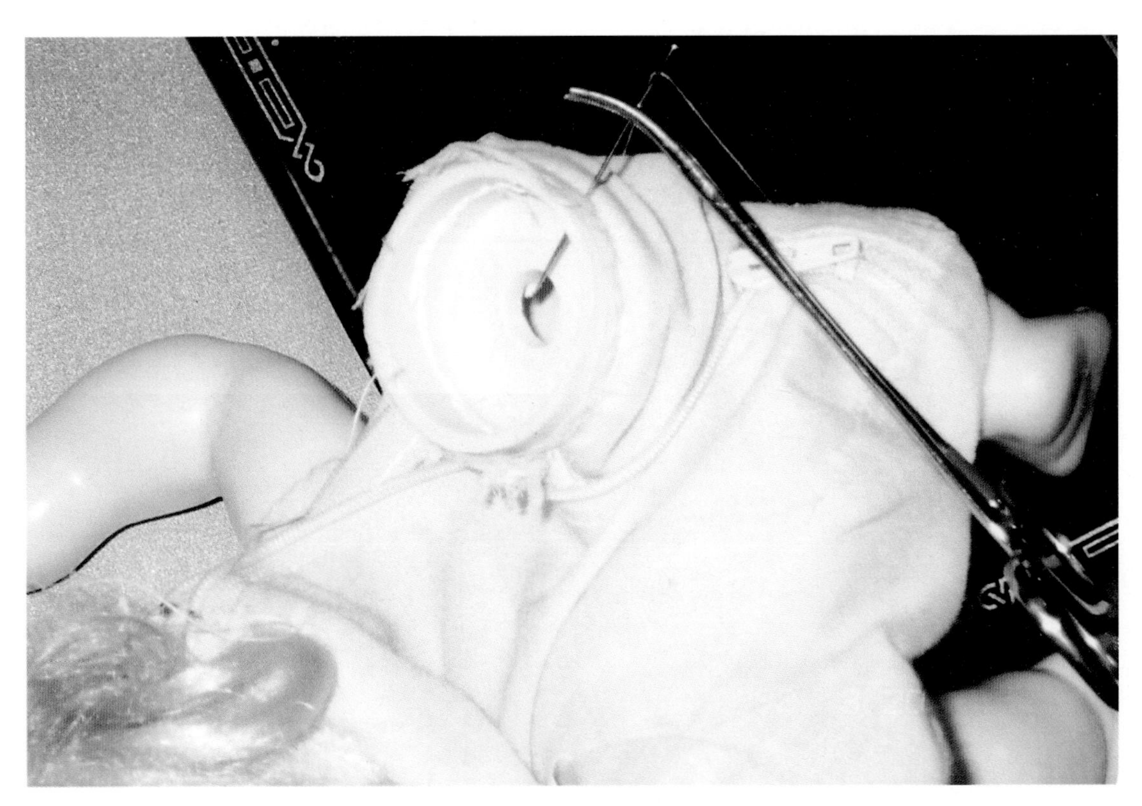

STEP #2. Sew the limb to the cloth body by hand..

Reattaching Limbs with Wire or Plastic Cable Ties

A small percentage of vinyl doll limbs and most all composition limbs are attached to a cloth body with a piece of wire, or a plastic cable tie if it is a later doll. (SEE PHOTO BELOW.)

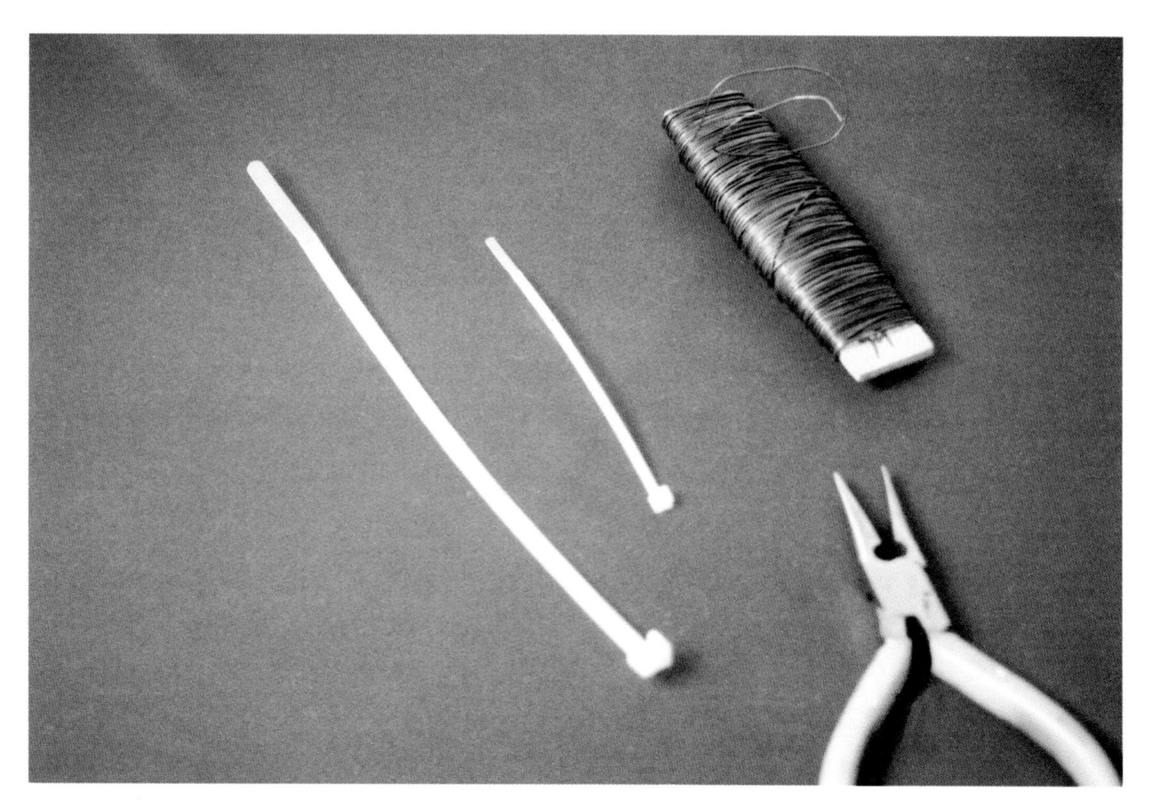

Cable ties and wire.

1 To take the doll's head off, find the wires that hold her head on. These wires are generally found on one of the doll's shoulders at the neck. (SEE #1A & 1B.)

2 Retrieve the ends of the wires from inside the cloth (slit it if you have to) and twist them until they open.

3 Take her head off.

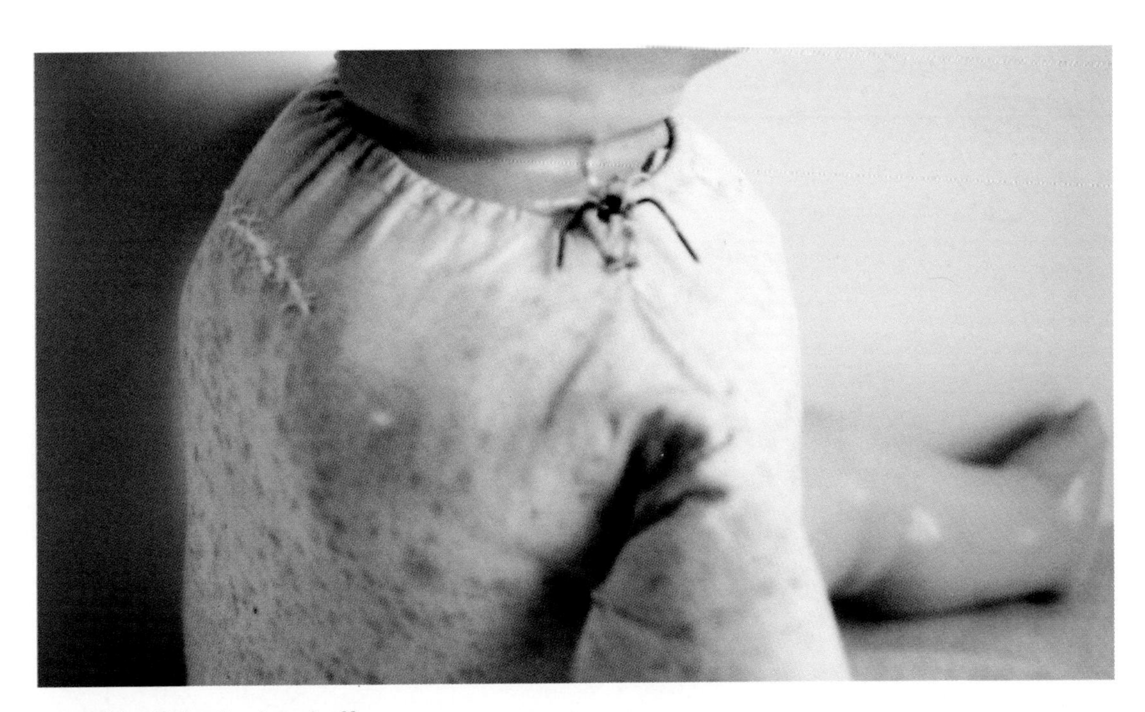

STEP #1A. Take the head off.

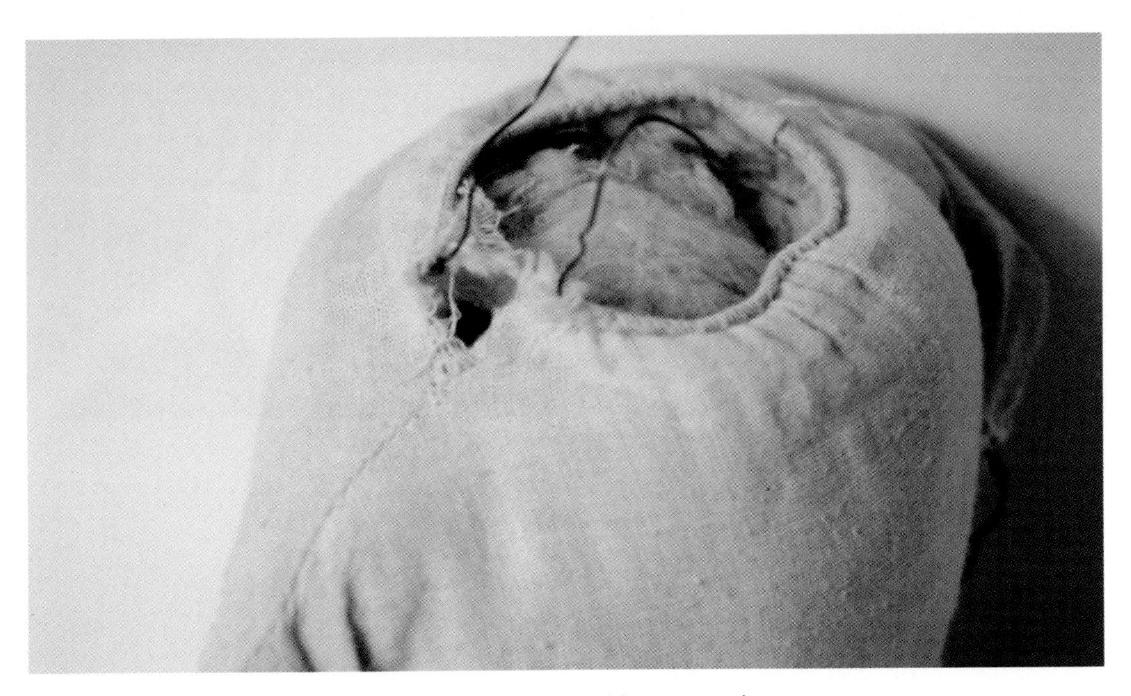

STEP #1B. The neck area of the body after the head is removed.

4 Note that the neck is covered with cloth across the hole. This is to prevent her stuffing from escaping her body and going up inside her head! If that happened, it would cause her body to become very limp and/or ruin the eyes, making them unable to sleep. (SEE **#4.**)

5 Remove her stuffing to the point that you can reach the affected arm or leg and be able to reattach the limb. (SEE **#5.**)

6 There is a groove in the leg or arm in which the wire or cable tie should be placed.

7 Insert the wire or cable tie into the proper casing in the cloth.

LEFT: STEP #4.
There is a cloth across the neck opening to prevent the stuffing from entering the head.

BELOW: STEP #5.
Removing the stuffing from the body.

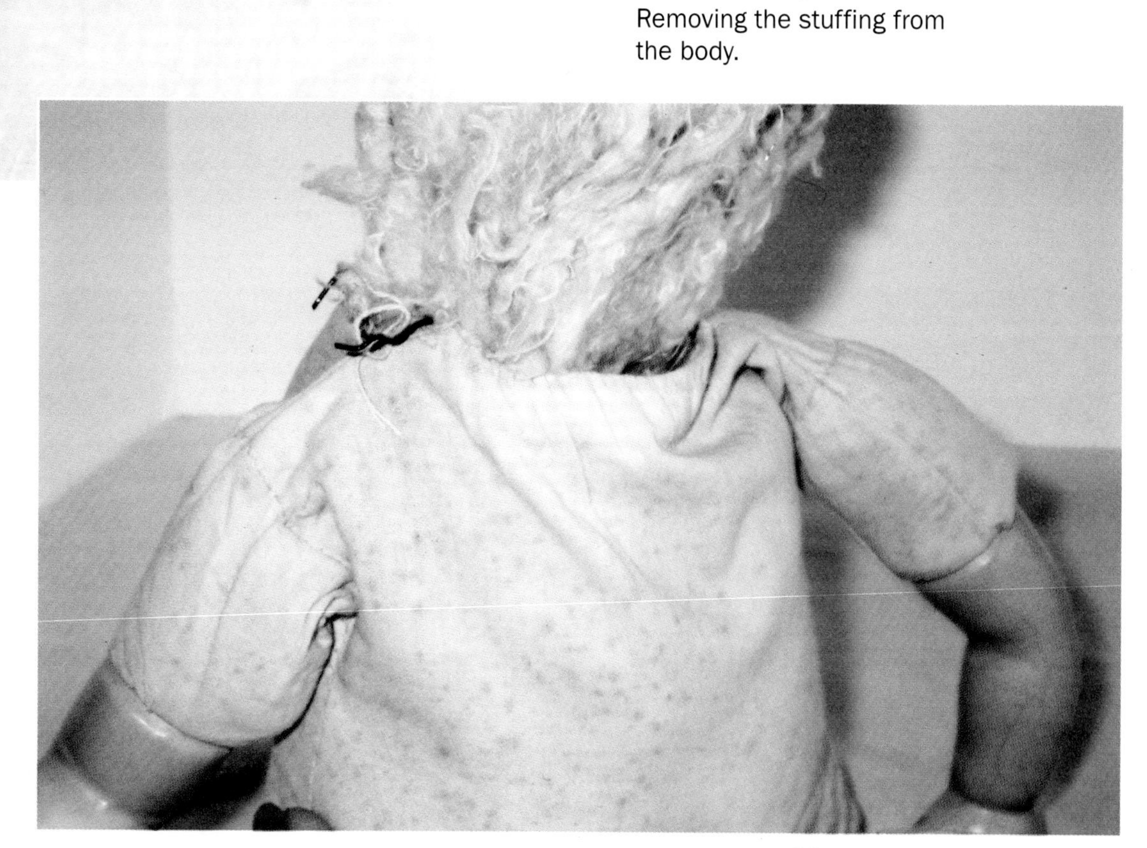

8 Make sure that it is positioned in the groove of the limb. (SEE #8.)

9 Twist the wire tight or pull the cable tie up tight.

STEP #8. The composition arm is attached with wire to the cloth body.

Now, when you are through stuffing her and putting her head back on, turn her to look at you.

We now have plastic cable ties that are used on all the new and modern dolls. The cable tie must be slim enough to lay properly in the groove around the leg or arm. Some cable ties are quite wide and thick. You will need slim ones to use on smaller dolls as they have smaller grooves. You can purchase assorted sizes from home repair stores or craft supply stores. Before you go shopping for them, measure around the leg and arm to make sure you purchase the ones that are long enough to not only fit around the limb but also to be able to tighten them in place. If they slip loose after you have the doll all put back together, you will have to start over. The cable ties have to be cut and thrown away, as they cannot be used again. Most professional doll hospitals do not use the new cable ties on old dolls as they prefer to use wire as was used originally.

If you are using wire (mostly on old dolls with composition limbs), you generally need to replace the wire that was used originally, as it is old and breaks easily when it is twisted tight. If this happens, you can sometimes retrieve the ends of the wire enough to "piece" together a new section of wire and twist it on to make it long enough for the job. This will save you from remodeling the cloth body at the neck by having to make a new casing for it. Purchase a paddle of wire at a hardware store about the same size as the old wire. Your restoration job will look more professional if you put wire back into a doll that had wire rather than using a new plastic cable tie in an old doll.

Many dolls have their heads attached with either cable ties or wire. They are usually run through a casing at the neck of the body. (SEE PHOTO BELOW.) If you are making a casing for a new body or an old one, use fabric on the bias or bias tape. You may want to tea dye it to match the color of the body as much as possible so it blends with the old fabric.

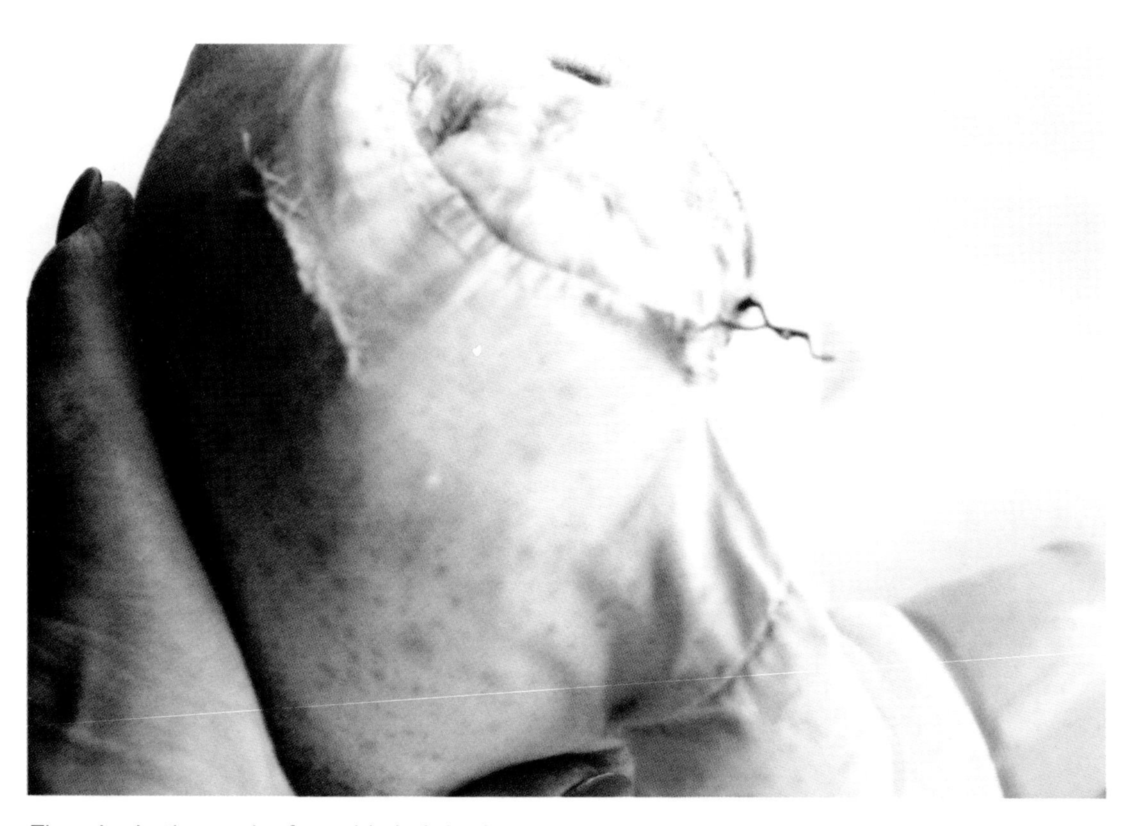

The wire in the neck of an old cloth body

Repairing a Very Old Cloth Body

If your vintage doll has a cloth body and her arms and legs seem to be attached with cloth folded under where the limbs connect to the body, then you have a chore ahead of you. These bodies were stuffed very tightly. It will be difficult to rip it apart and remake a body that will look and feel like the original. The stuffing is usually a gray or black cotton type material which, when you try to remove it, comes out in small pieces with each tug about the size of a teaspoon. These bodies, however, do need repair – usually around the neck or where the arms and legs are joined. Iron-on tape is the magic answer which will make the repair look as professional as possible.
(See photo below.)

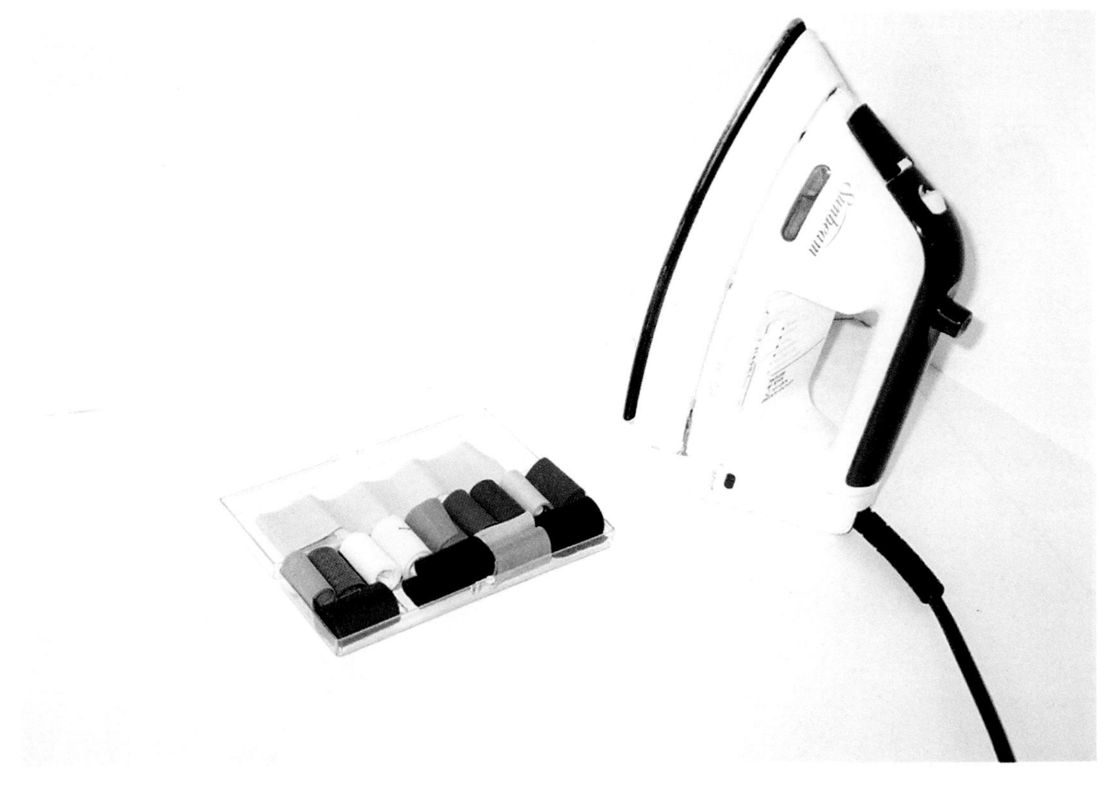

Iron-on tape.

1. Choose a piece of iron-on tape that is as near the color of the body as possible. (SEE #1.)

2. You may need to put more stuffing into the place where the body is torn in order to make a good "pad" on which the iron can rest to make the tape stick well. (SEE #2.)

3. Cut the patch round or oval without corners. Corners tend to get caught and pull up easily, destroying your hard work.

4. Heat the iron and hold it on the tape for a few seconds without moving it.

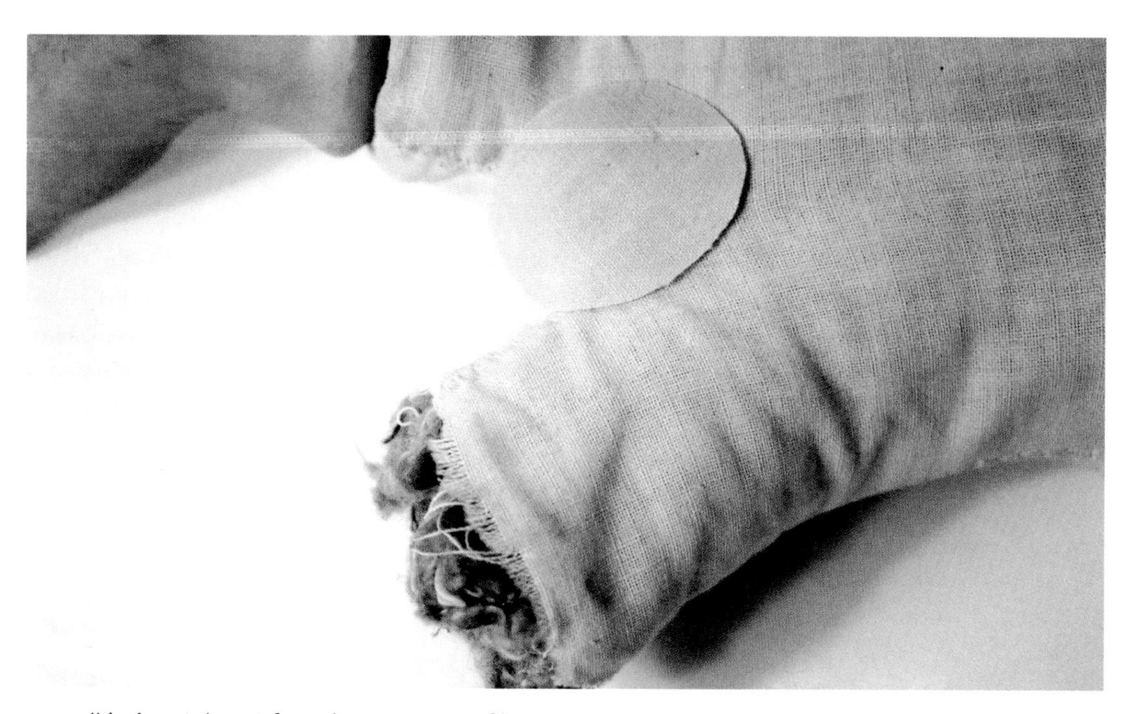

STEP #1. A patch cut from iron-on tape. Choose a tape color to match the color of the body.

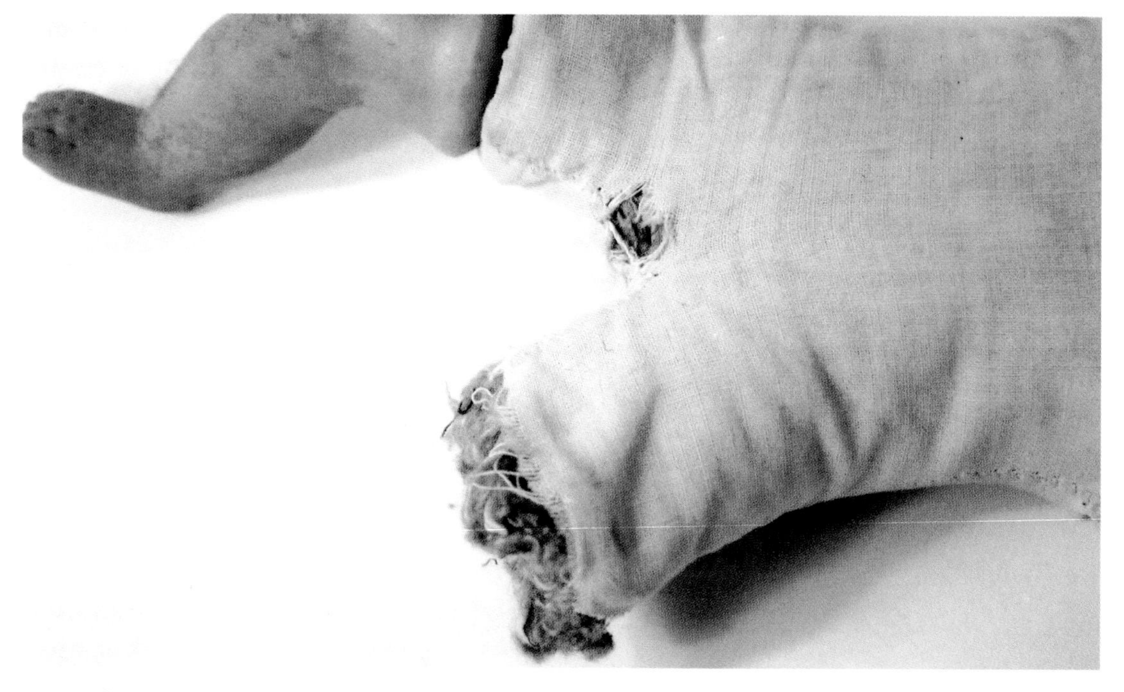

STEP #2. A hole in the cloth body at the crotch.

5 Use the point of the iron (with some pressure applied) around the edges of the patch so it makes a good seal. This strengthens the old fabric of the body, and holds it all back into the original shape as well. (SEE #5.)

NOTE It is always a good idea to mention to the customer that this is the way you may have to repair the body. Explain that sometimes the fabric is just too old to hold stitches when you are trying to pull the tear or hole together. They need to understand what you are trying to do so they do not think you are using a modern method to fix their old doll. If they do not want the tape to show, but want the strength of it, then you can cut patches of old material that matches the body as best you can and appliqué a patch over the iron-on tape, so it will not show at all.

6 If you had to remove an arm or leg in order to make this repair, it was probably held in place with fine wire.

7 Find the wire ends inside the cloth by feeling around the leg or arm where the cloth meets the limb. (SEE #7.)

8 Make a very small slit in the fabric to retrieve the wire ends and perform the task.

9 Be gentle with the wire as you may reuse it to reattach the limb.

10 Put the ends of the wire back through the slit to hide them. Then you can easily repair the small slit with a needle and thread.

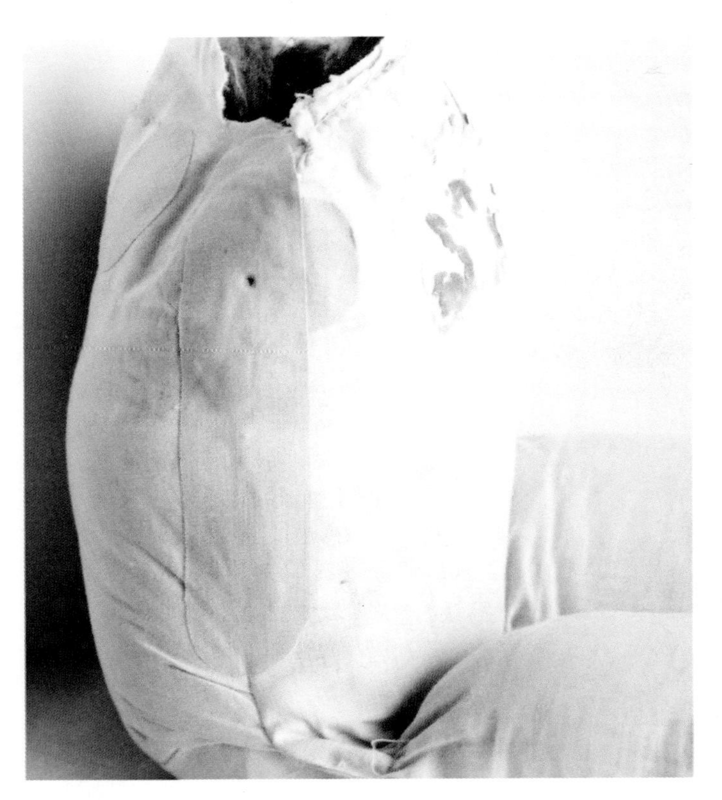

STEP #5. Lots of iron-on tape is used to repair and strengthen an old cloth body.

STEP #7. Find the wire in the leg of the cloth body.

Doll Eyes

Cleaning a Doll's Eyes

1 Place the doll face up on a towel on your lap.

2 Hold her eyes open with your finger.

3 Dip a cotton swab in the cleaner and saturate her eyes with it. Do this while they are open and then again when you allow them to close.

4 Make sure it is thoroughly wet and swab her eyes. (SEE **#4.**) It will not hurt if the cleaner goes inside her head.

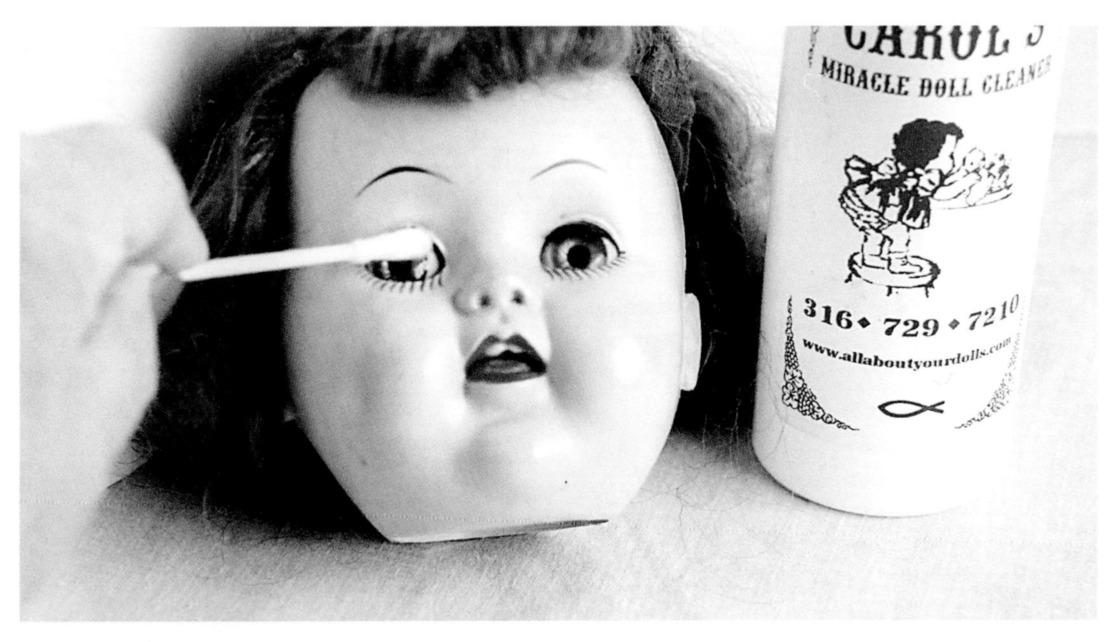

STEP #4. Put cleaner in the eyes with a soaked cotton swab.

5 When they are really wet with the cleaner, use a wooden toothpick (SEE #5.) and rub the entire eye, even in the crevice, all the way around. Repeat this process, if necessary, until everything has been cleaned out.

6 **Note:** Many times, a doll's eyes are all full of white residue. Mildew also makes the eyes look cloudy and this same treatment will clean mildew as well. My cleaner can be used on all kinds of eyes — acrylic, vinyl, tin and glass. (I have never had a problem cleaning eyes on any doll.)

7 You may also put a drop of sewing machine oil or 3-in-1 oil in each eye to keep them working freely and to brighten the color. (SEE #7.) Since there is a lot of controversy over putting oil in dolls' eyes, you should follow this step at your own risk. I can say that I have used it myself for years and my customers have been very happy with the results.

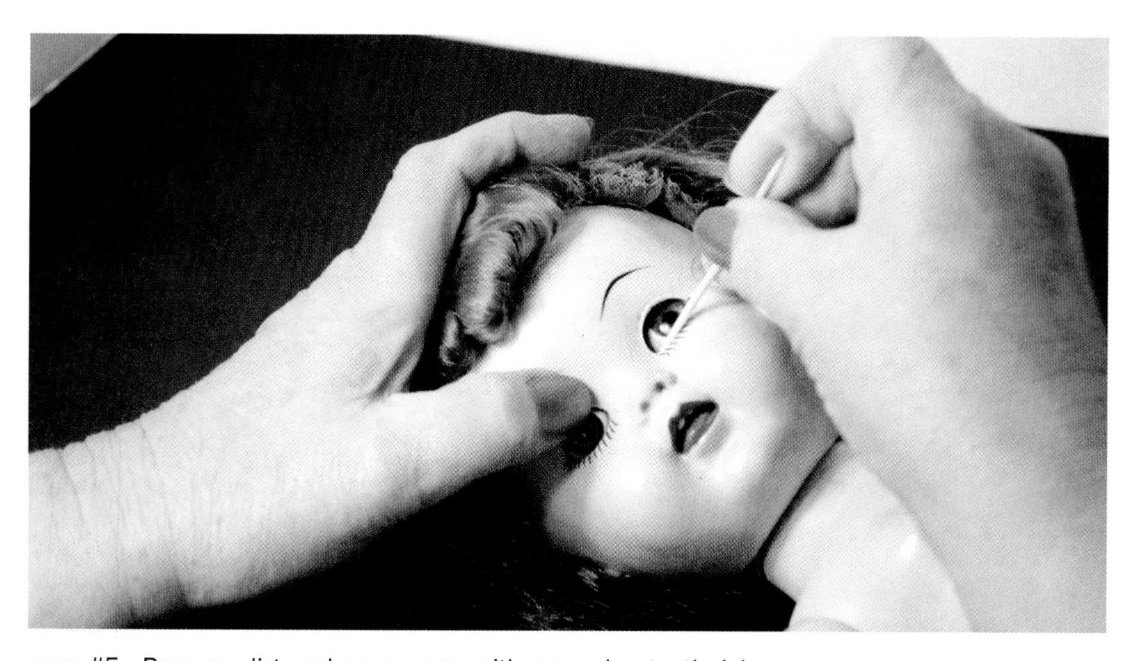

STEP #5. Remove dirt and soap scum with a wooden toothpick.

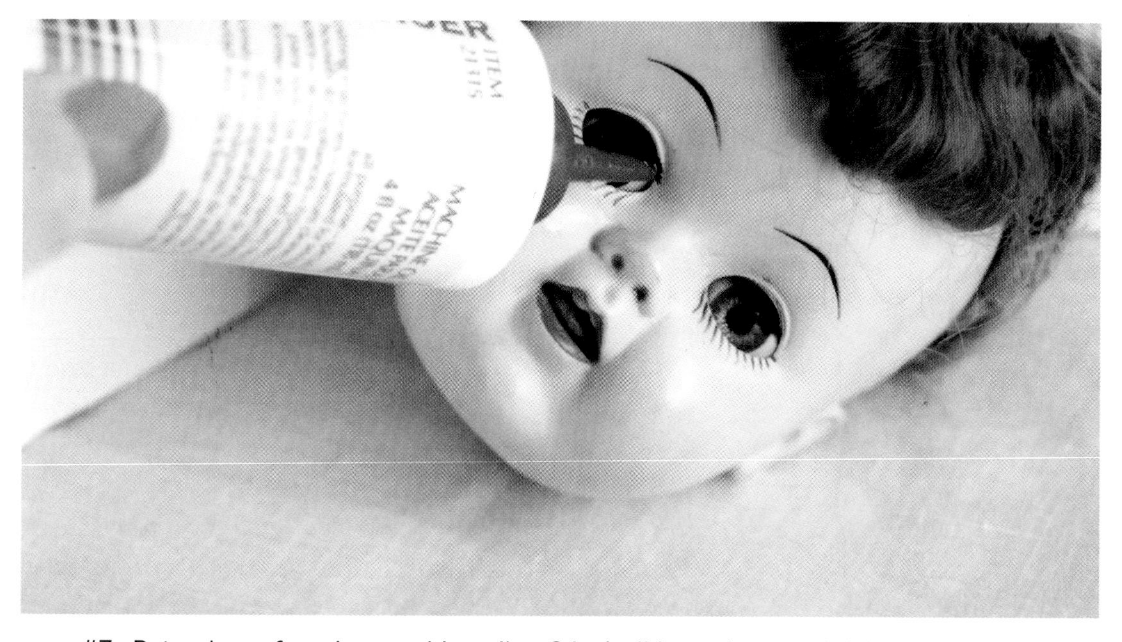

STEP #7. Put a drop of sewing machine oil or 3-in-1 oil in each eye to brighten them.

Restoring Cloudy or Shattered Glass Eyes

Cloudy or shattered eyes on old composition dolls are usually glass. They look as if they have shattered but all the pieces are still in place. (**SEE TOP LEFT PHOTO BELOW.**)

1. Lay the doll down face up on your lap and place a few drops of sewing machine oil or 3-in-1 oil into each eye. (**SEE # 1.**)
2. Lay her aside face down for two or more hours. The iris (colored part) of the eye will return, but the pupil (black center) will not.
3. Use a cotton swab and carefully wipe the oil off.
4. Lay her on her back again.
5. Hold her eyes open.
6. Use a felt-tipped permanent pen to make a pupil in the center of each eye. (**SEE #6.**)

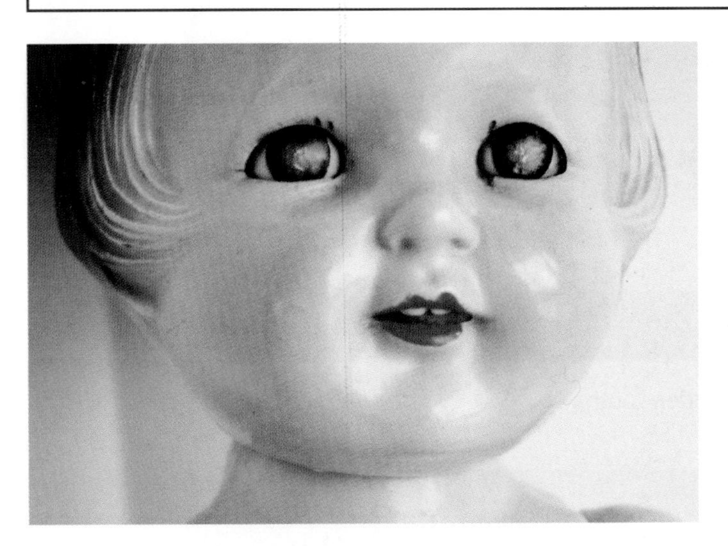

Glass eyes that have shattered.

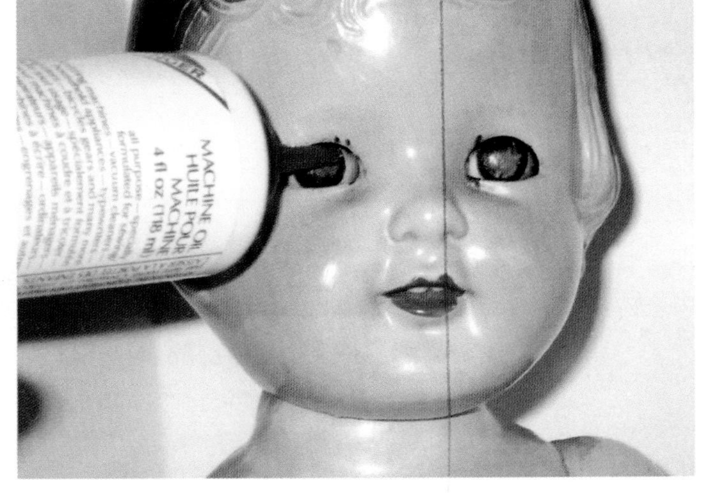

STEP #1. Put a drop of sewing machine oil or 3-in-1 oil in each eye.

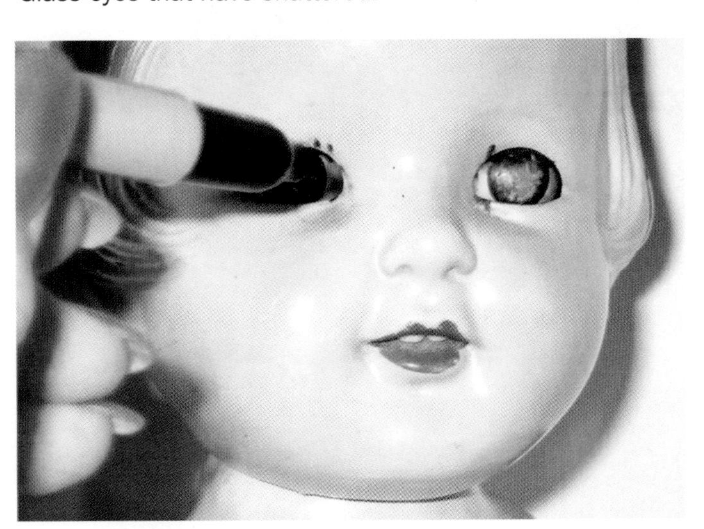

STEP #6. Make a pupil in each eye.

TIP There are paints (on a card similar to water colors) that are available at hobby or craft stores. I believe they are made to use in stained glass crafts. This paint is translucent. When you use the original color over a glass eye that has been crazed, the eye comes alive! For instance, use blue on a blue eye. Try this on one of your "practice patients" and see for yourself.

Replacing Eyelashes

1 Lay the doll on her back.

2 Select an eyelash that is as close to the original one in color and thickness as you can and have it ready.

3 Have glue ready on a piece of paper and a toothpick beside it.

4 When one eye is open, both will open. Use the good eye to hold both eyes open. (**SEE #4.**) On the bad eye, you will see a slit where the eyelashes used to be.

5 Open the slit as best as you can in order to put the new eyelash in place. A good tool to use is a flat tool with a sharp point that is used to clean ceramics. An X-acto knife also works well.

6 Run a corsage pin (a little bigger and longer to hold onto than a straight pin) in the slit to make sure you have it open all the way across the eyelid. (**SEE #6.**)

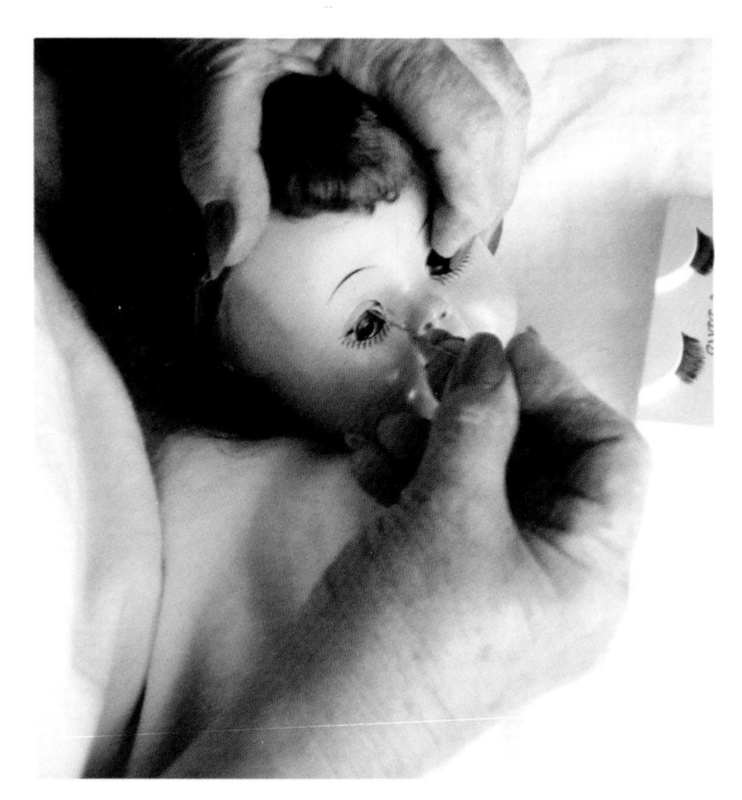

STEP #4. Look for a slit in the eye where the eyelashes should go. Open the slit in the eye and apply glue using a toothpick.

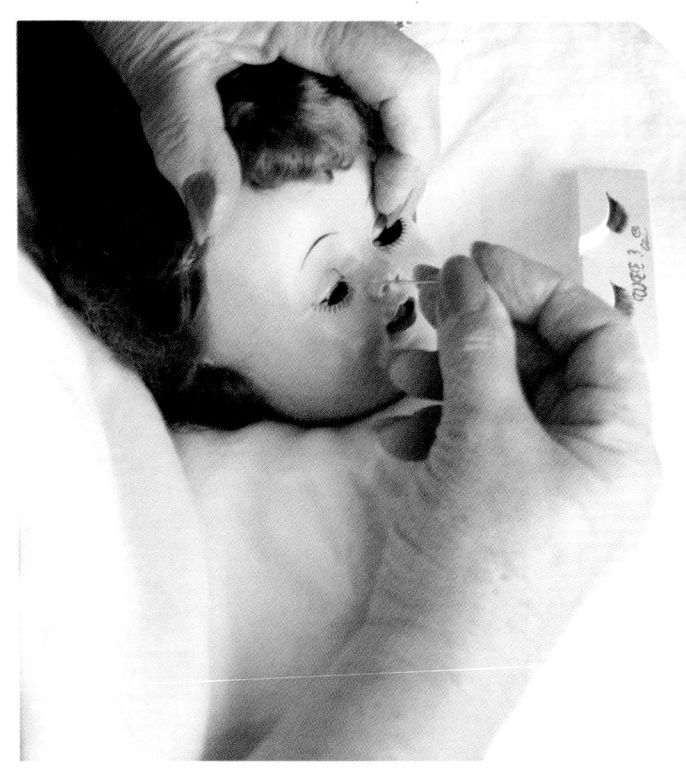

STEP #6.

7 Dip a toothpick in the glue (Aleene's Tacky Glue is recommended).

8 Carefully run a thread of glue right across the slit. (SEE **#8.**)

9 Place the eyelashes on the glue and use the pin to poke them into the slit. (SEE **#9.**)

10 Go along the entire length of the eyelid to be sure the eyelashes adhere well.

11 Let them dry thoroughly before you let her eyes close. You can use a clean toothpick to put down in the crevice beside the other eye to hold the eyes open, while it dries. Leave her lying down until the glue has time to dry.

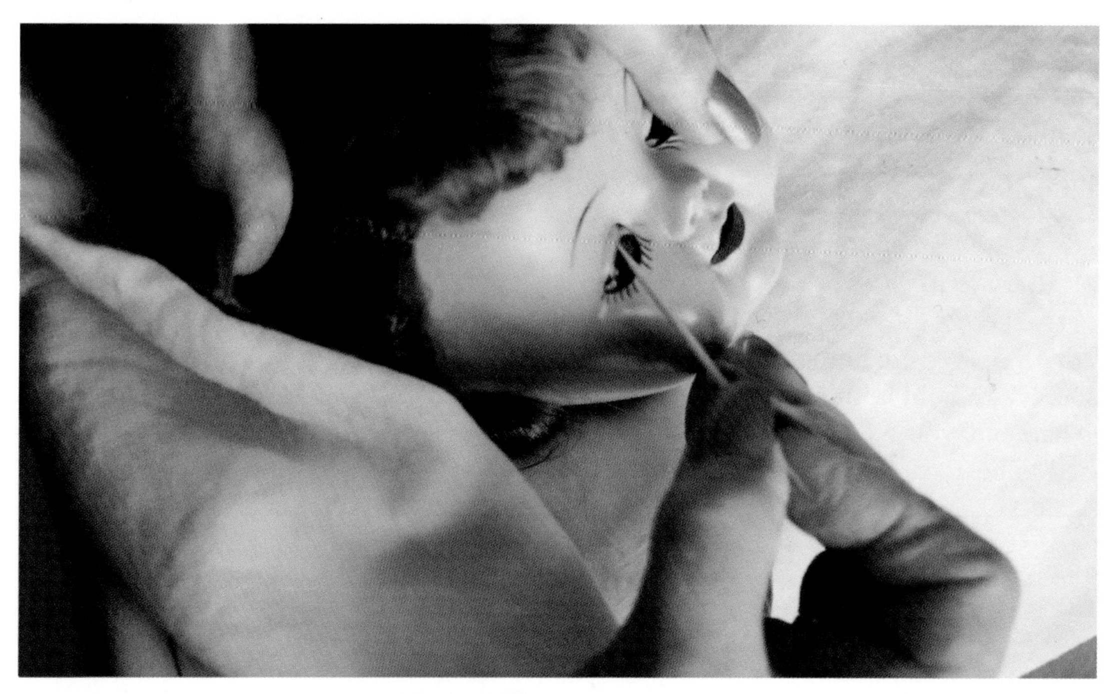

STEP #8.

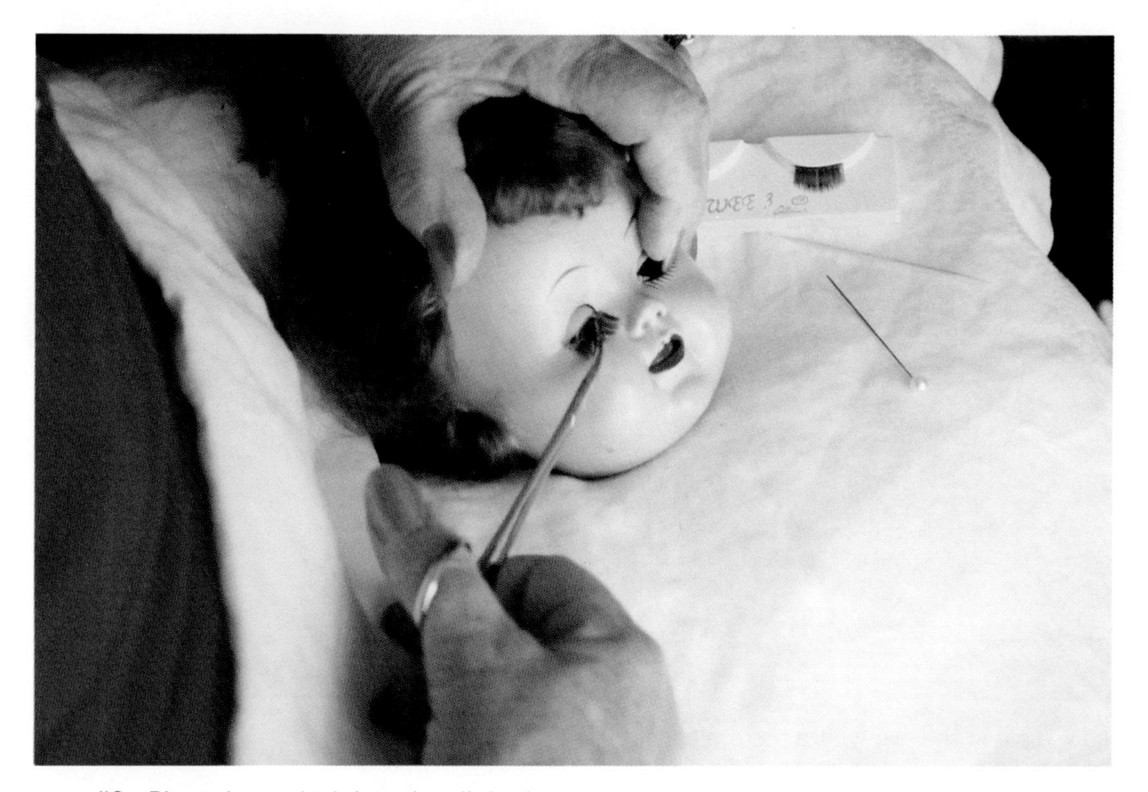

STEP #9. Place the eyelash into the slit in the eye.

Replacing or Resetting Glass Eyes

During the many years that I restored dolls, I was blessed to have a friend that mastered the art of resetting glass sleep eyes. While I watched the procedure, I never attempted to learn it and, therefore, it will not be included here as this friend is still available.

If you are interested in pursuing this phase of doll repair, I recommend an out-of-print book — *Make, Repair and Restore Eyes in Bisque, Composition, Vinyl and Plastic Dolls*, printed by Aurore Publishing, P. O. Box 752, Worthington, Minnesota 56187. The book, with wonderful illustrations, covers everything you will ever need to know about doll eyes. If you are able to find a copy, it will be well worth the trouble.

Fixing Doll Eyes That Are in Separate "Pockets" in the Head (Vinyl Doll)

For repairing the eyes of vinyl dolls which are in separate "pockets" in the head, the following instructions are given.

1. Use a hair dryer to heat that portion of her head and the eye.
2. Squeeze her head from the back.
3. Push hard against the eye sockets from the back with your thumbs.
4. Do whatever you have to do to pop the eye out of the front of her face.
5. Clean the eyes using the doll cleaner.
6. If the eyelashes need to be replaced, follow the previously detailed instructions in the section on "Replacing Eyelashes" in this chapter.
7. When you are finished, heat the head again and push the eye back into the socket from the front of her face.

Resetting Stationary Eyes

If you need to reset or replace glass eyes that do not sleep, hopefully you will have one or part of one that was original to the doll. This helps in determining the size of the pupil, and the eye itself.

Glass eyes of this sort are almost always in bisque head dolls. The open dome dolls give you plenty of room in which to work. Pink setting wax is used around the inside of the eyehole in the head.

This method is for eyes that **do not** sleep.

1. Place the eye against the wax from the inside of the head. The eye will stay in place while you prepare to adhere it to the head.

2. You may now prepare some plaster of paris according to package instructions.

3. Put a dollop on the outside of each eye to hold them in place.

4. Lay the doll down on her face on a padded area and let the plaster of paris dry thoroughly.

TIP Do not use the acrylic eyes, which are now available, in an old bisque head. Glass eyes are available from doll supply sources

Doll Clothing

Restoring Doll Clothing

Laundering

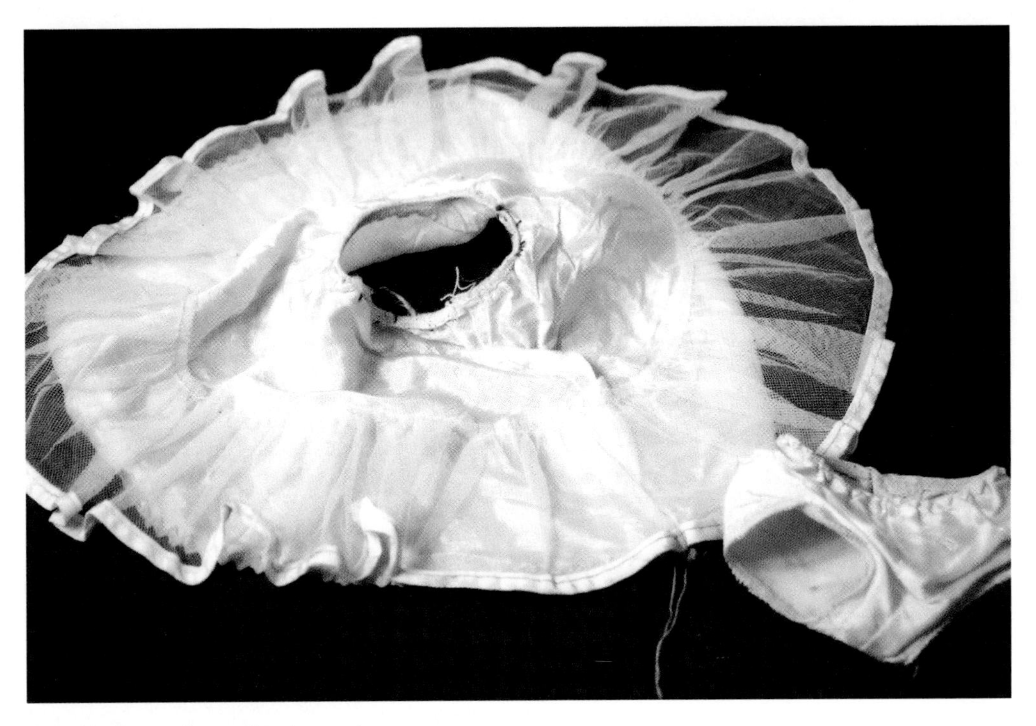

A half-slip and panties to restore.

A dress to restore.

1 Take the clothes off your doll carefully, making sure they are not damaged further by removal.

2 Examine each piece of clothing closely and determine the type of fabric. Cotton fabric can be washed while wool or taffeta should not be washed. Most clothing on a doll older than those from the 1960s will be made of cotton or perhaps satin, taffeta, silk, net or tulle, if it is a formal outfit such as a bride. (The original *Miss Revlon* dress shown in the illustration is cotton, while the half-slip and panties shown are silky taffeta with nylon net trim.)

3 Before cleaning of any kind is done, turn the item wrong-side-out. There are probably many ravels of fabric on the seams as well as long threads left when the garment was made. Trim them carefully paying special attention to only cut away the excess of ravels and threads and not the seams themselves! (SEE #3.)

4 Note if there are buttons or snaps missing and replace them at this time.

5 Ripped or pulled open seams and loose hems are easily repaired by hand at this time.

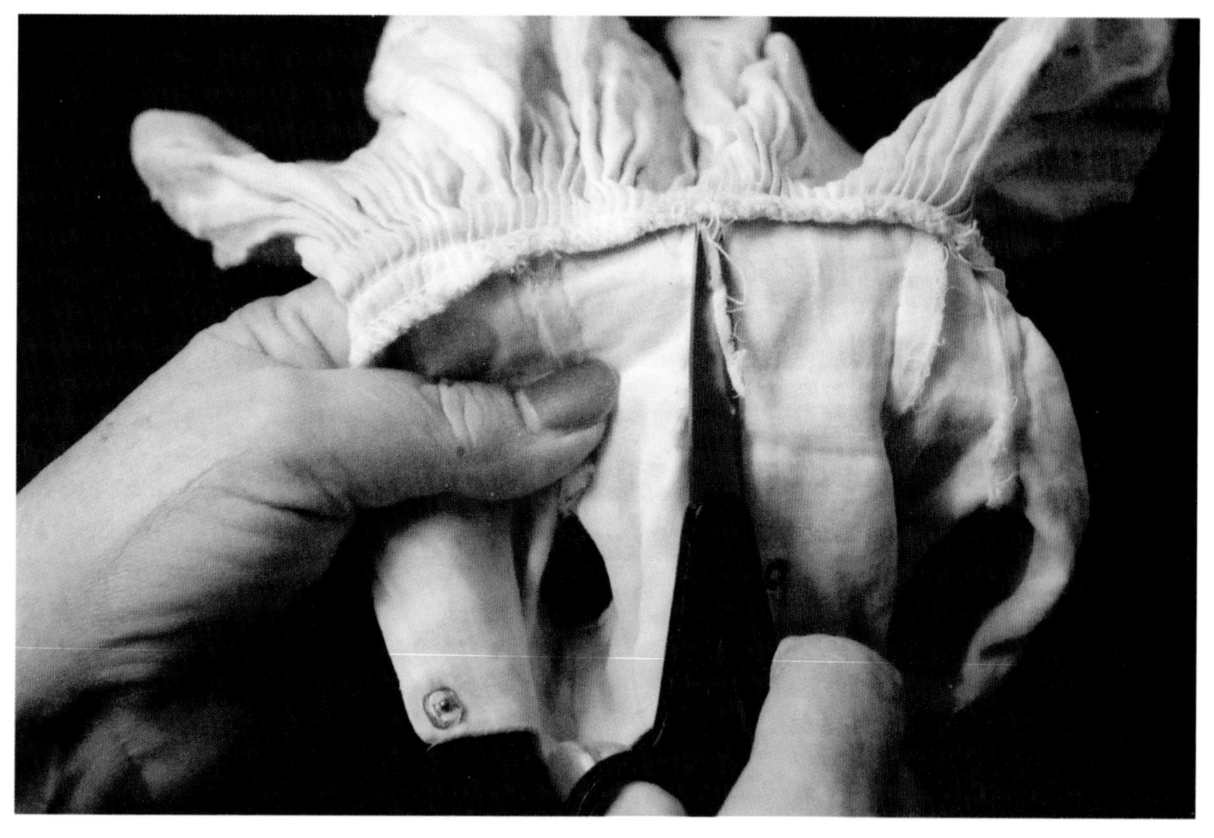

STEP #3. Trim the ravelings and threads from inside the dress.

6 Stretched out elastic in sleeves or waistbands should be removed at this point and replaced after laundering and/or ironing. (SEE #6A & #6B.)

7 Turn the garments back to right-side-out.

8 Determine if the fabric can be washed or not. You may have a few small pieces of fabric you just trimmed that you can wet (if nothing more than to see if it fades).

9 Prepare warm water with soap of some kind. (Palmolive® liquid dishwashing soap is recommended.)

STEP #6A. Remove the old elastic from the waist of the panties.

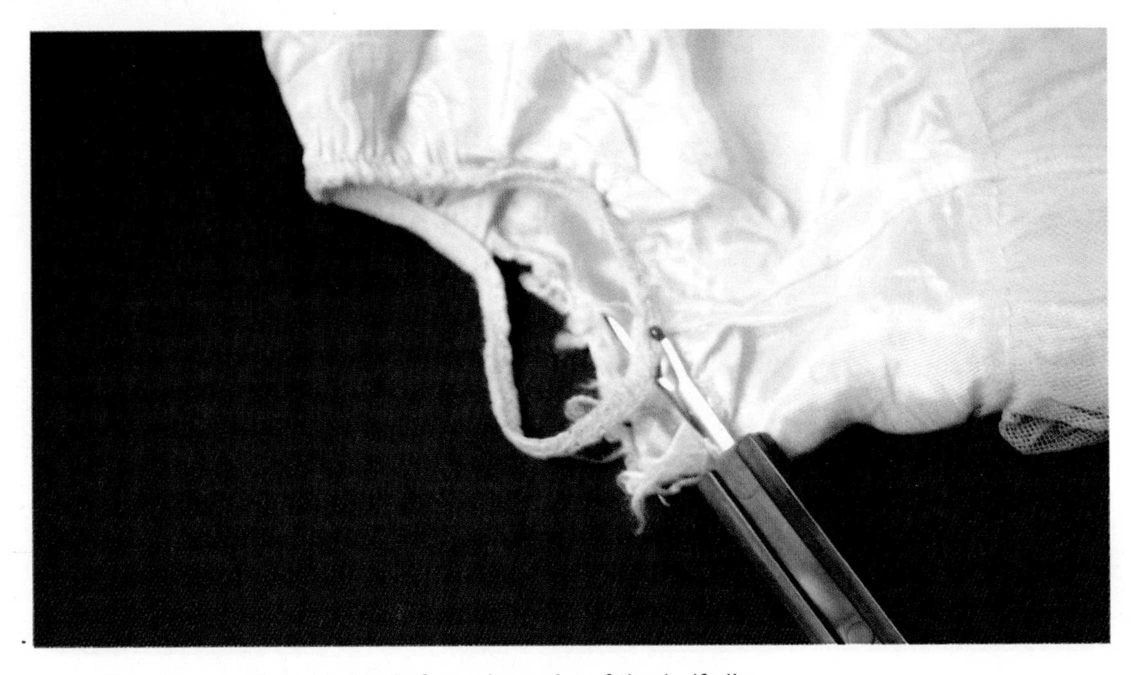

STEP #6B. Remove the old elastic from the waist of the half-slip.

10 Submerge the garment in the warm water. (SEE #10.)

11 Leave it soaking for a few minutes. If the garment is very soiled, it may take more than one session.

12 When the garment is clean (but perhaps not really as bright in color as you think it should be) remove the garment and add liquid beach very sparingly, a spoonful at a time to the water. Swirl the water to mix well before putting the garment back in to launder. (A two-pound margarine tub is recommended for this process.)

13 Submerge the garment in the bleach water and keep stirring.

14 The colors will brighten and even patterned fabric will look brighter with this process. (SEE #14.)

15 Rinse several times. Be sure to get all the soap out of the garment.

16 Roll in a towel.

17 Squeeze very well. This will dry the garment so that you need not hang it as the weight of the wet garment might pull it out of shape.

18 Sometimes the article of clothing will fit right over a drinking glass or something else handy in your kitchen! It will dry with fewer wrinkles and also air can circulate around it. If not, just lay it on a clean towel to dry.

STEP #10. Wash and bleach the dress.

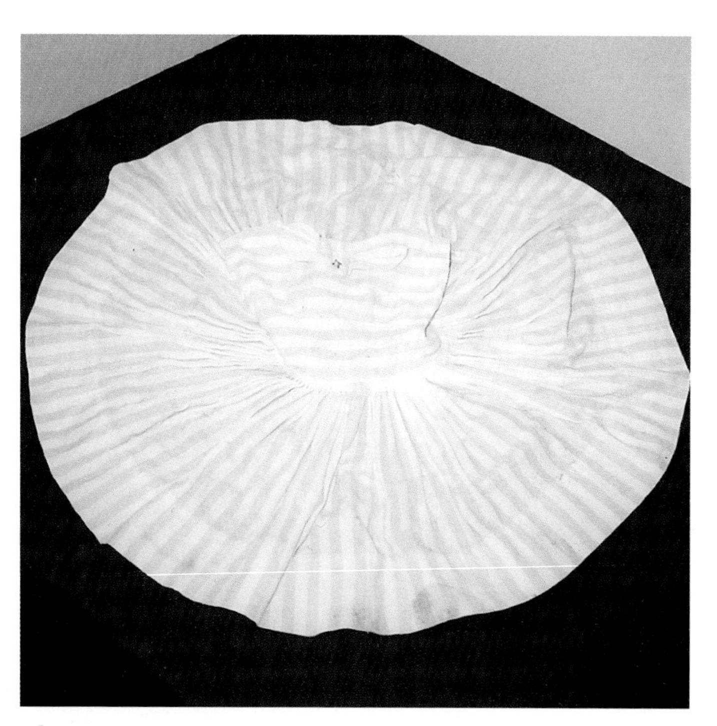

STEP #14. The dress is clean and bright.

Ironing

In addition to an ironing board, you will find a sleeve board, a steam iron, a puff iron and spray starch will make ironing the laundered doll clothing much easier.
(SEE PHOTOS ON THIS PAGE.)

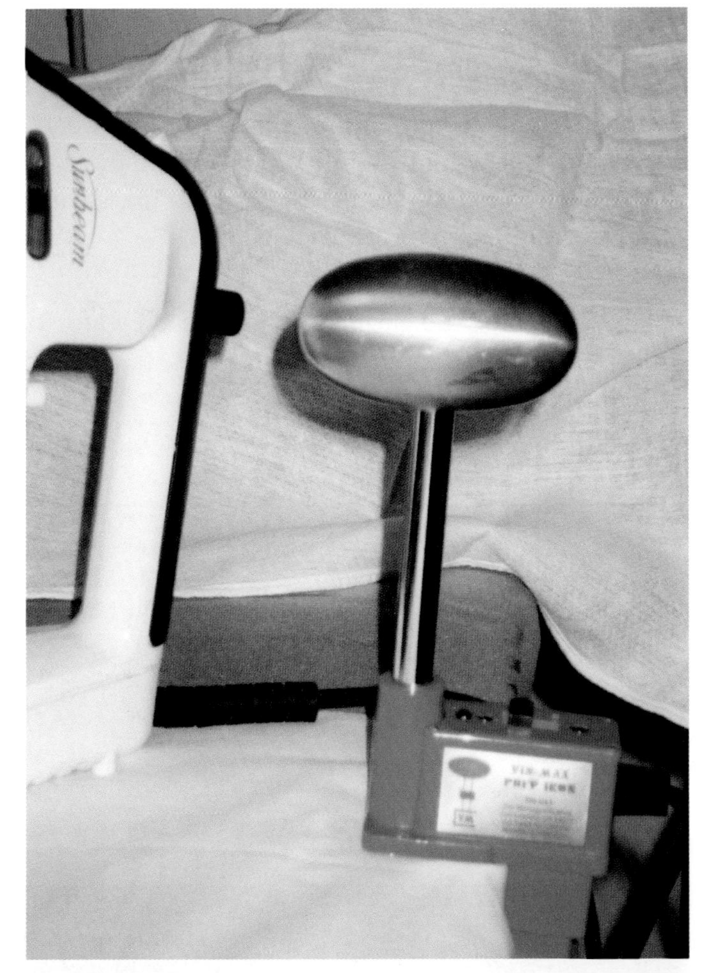

RIGHT: "Puff iron."

BELOW: Ironing tools.

1 Lay the garment down on your ironing board and spray the whole garment generously with the spray starch. (SEE **#1.**)

2 Bundle it up and hold it just a few seconds so that the starch can be absorbed evenly throughout. (SEE **#2.**)

3 Set your iron on the correct temperature (or just a little cooler) for the fabric.

STEP #1. Spray starch all over the dress well.

STEP #2. Wad the dress up and hold it to distribute the starch well.

4 Take care to place each area of the garment on your sleeve board for maximum access with the iron. (SEE #4.)

5 If you have a dry spot or a crease pressed into the garment that you do not want, spray a little starch onto your finger – instead of on the garment – and apply it only where you want it. (SEE #5.)

STEP #4. Place the dress on the sleeve board for ironing.

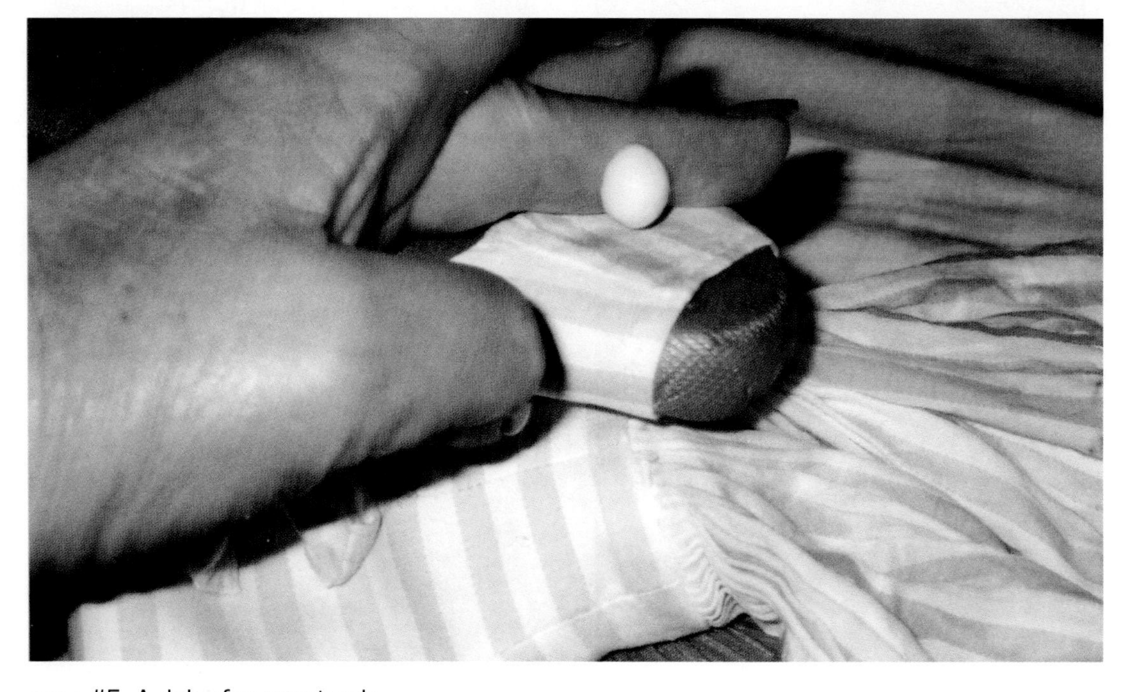

STEP #5. A dab of spray starch.

6 Move the garment on the sleeve board so the iron has access to every fiber of the fabric. (SEE #6.) Be sure to take your time and let the iron stay in an area long enough to completely dry that area. The fabric will hold the smooth look for which you are striving.

7 Take care to iron up into the gathers with the point of the iron. (SEE #7.)

8 If there is lace trim and it is gathered into a ruffle, then it should be ironed last, with the point of the iron going up into the gathers so that each and every area of the ruffle has been touched with the iron.

9 Move the garment to the more spacious part of your larger ironing board to make the circular motions around the ruffled lace.

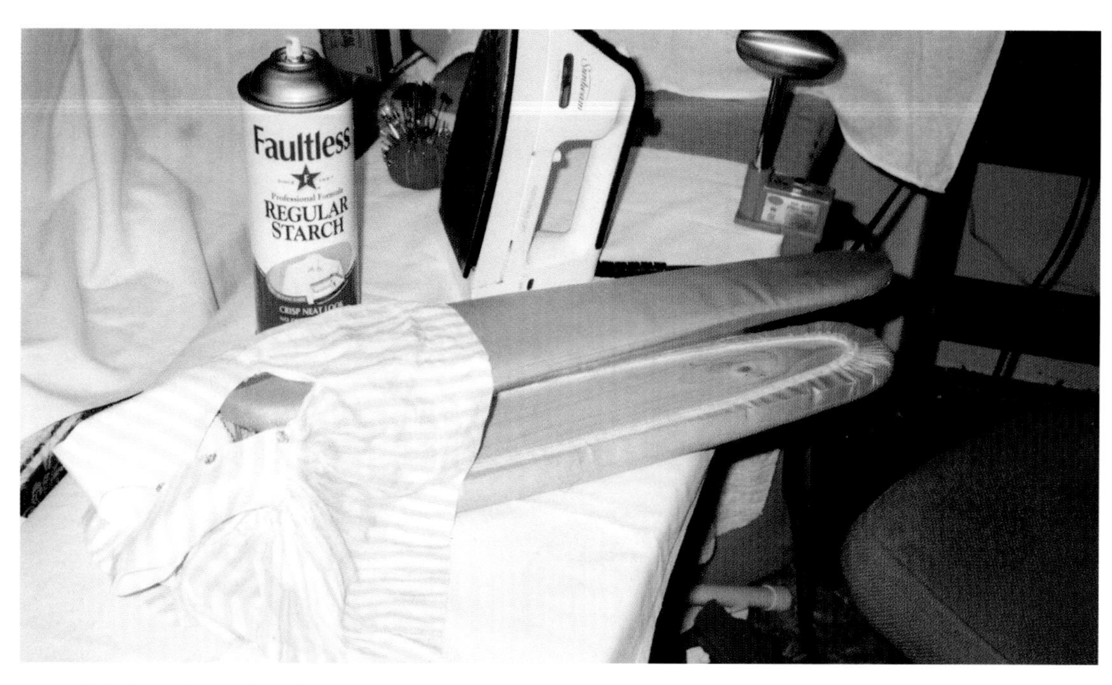

STEP #6. Iron the skirt of the dress.

STEP #7. Iron up into all the gathers.

10 For fabrics such as taffeta, silk, nylon net, satin or any other questionable fabric, be sure to test the ironing technique on a small hidden area before proceeding. (SEE **#10A & #10B.**)

11 Some items need to be dry-cleaned rather than washed. If you have a garment professionally dry-cleaned, it is recommended you ask them **not** to press the garment.

12 Take the cleaned garment to your sleeve board with a warm (not hot) iron. Take your time pressing! Many times this must be done on the inside, depending on the fabric.

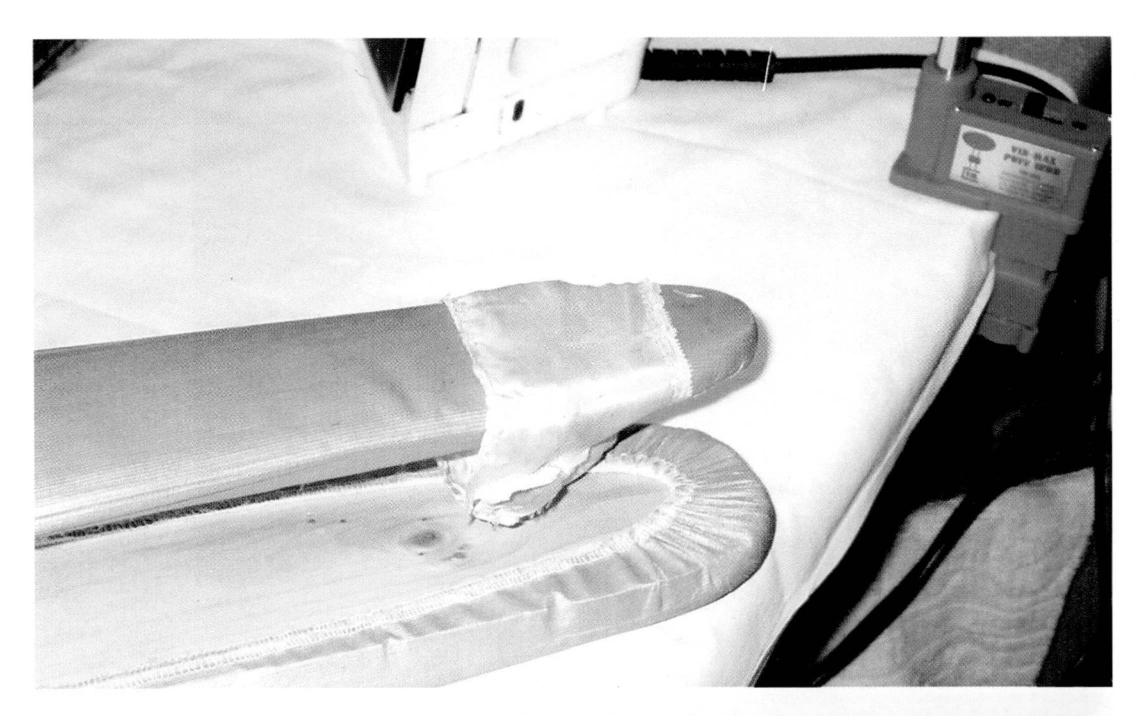

STEP #10A.
The panties are ready to iron.

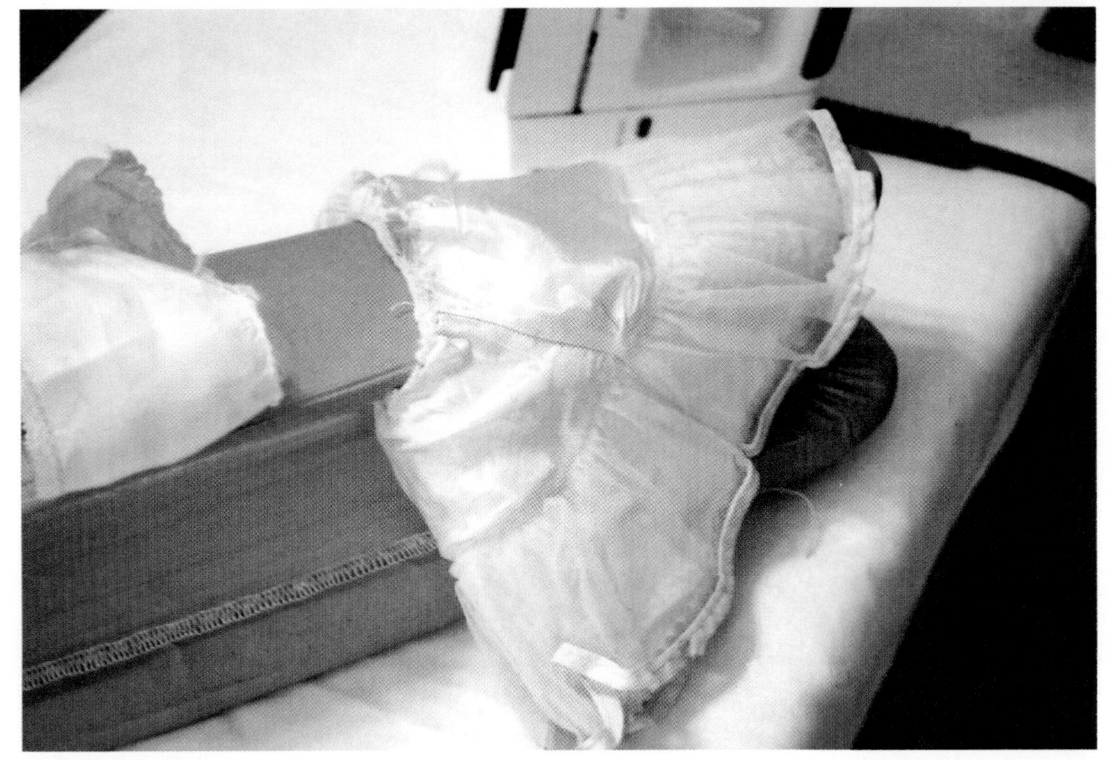

STEP #10B.
The half-slip is ready to iron.

Mending

Iron-on tape, which comes in an assortment of colors, is excellent to use in mending doll clothing.

1 Cut a piece of iron-on tape just big enough to cover the hole.

2 Apply it on the underside of the garment.

3 If the hole has threads hanging from it, clip those off before using the tape. Obviously, the patch will show more on thin material than it will on cotton fabric, but it also strengthens the fabric.

Reproducing the Original Clothes

If you find that the original clothing on the doll is beyond restoration, the next best step to follow is to replicate the original clothing using the old garment for a pattern.

1 Use fabric as similar to the original as possible. A thrift store is sometimes a good place to find an item that has been washed many times but is still good strong fabric.

2 The replicated clothing should appear to be the same vintage as the doll and should look like the original as much as possible. In other words, an old doll should not be wearing a new brightly colored dress with synthetic lace and ribbon.

3 If you do a lot of sewing, you probably have a stash of old bias tape, rickrack and ribbon. (**SEE #3A & #3B.**) If not, start collecting these items. Estate sales are a wonderful place to find them. Since you will be making small items, you can use the leftover packages of trim and thread. Small buttons and snaps also should be saved.

4 Patterns for doll clothes are a find also. Do you know that all the major pattern companies made patterns especially for certain dolls while they were popular? Patterns for *Shirley Temple*, *Chatty Cathy*, *Ginny®*, *Miss Revlon* and *Barbie®*, as well as many others, were readily available. Look for these at doll shows and thrift stores. (**SEE #4.**) Reprints of many of these patterns are also available. Check the classified sections of doll magazines or search the Internet for them.

STEP #3A. Stash of all kinds of buttons and snaps.

STEP #3B. Stash of rickrack, bias tape, ribbon and trim.

STEP #4. Old patterns for doll clothing.

Restoring Doll Shoes

There are many kinds of doll shoes (old as well as newer) and they are made of many different materials. Old shoes are well worth the time it takes to restore them.

1 Closely examine the pair of shoes. Determine what you can do to make them look better. Sometimes just trimming the edges very carefully makes a big difference in their appearance. (SEE **#1.**)

2 Use a little dab of glue on the end of a toothpick to fix a strap or sole.

3 To clean them, dip a cotton swab in the doll cleaner and rub over the shoe. The dirt as well as most of the scuffmarks will disappear.

STEP #1. Trim ravelings from the doll shoes.

4 If the hole for a button, snap or shoestring is pulled out, a very small piece of iron-on tape on the underside will do wonders. (**SEE #4A & #4B.**) (White tape is used in the illustration so it can be more easily seen. However, you should use a color that matches or comes close to the color of the shoe.)

RIGHT: STEP #4A.
Use iron-on tape to mend the strap of the shoe.

BELOW: STEP #4B.

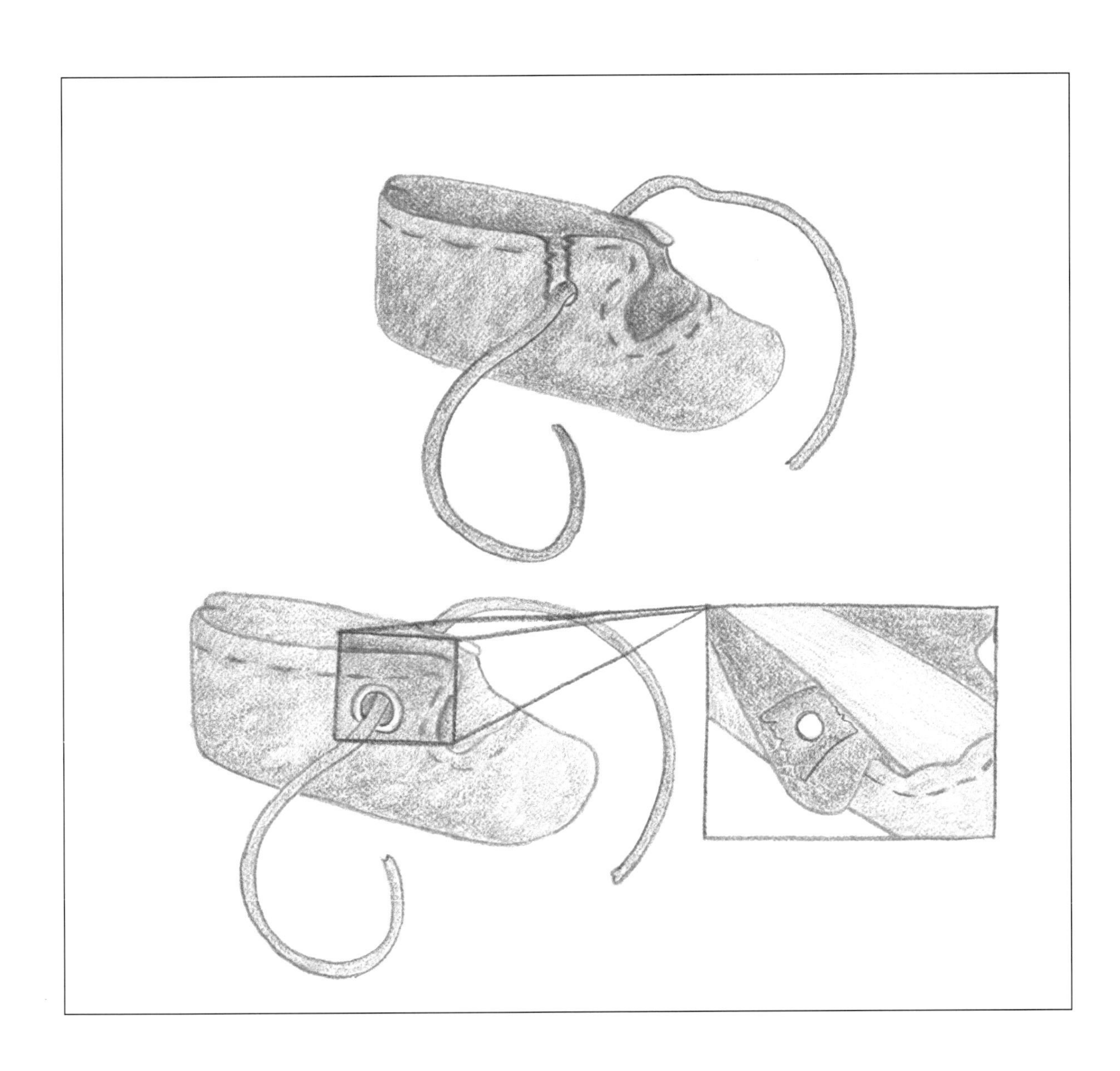

5 Make a hole through all the layers and the piece of tape will strengthen the torn area as well as hold it together. Furthermore, it is usually quite invisible from the right side. (SEE #5.)

6 Use adhesive tape or even the tape part of a Band-Aid® to line a strap which is falling apart.

7 Scuffs that are hard to cover on black or brown shoes can be made to look like new again by marking with a felt-tipped permanent pen in the appropriate color. Be sure to let the shoes dry well before putting them on the doll.

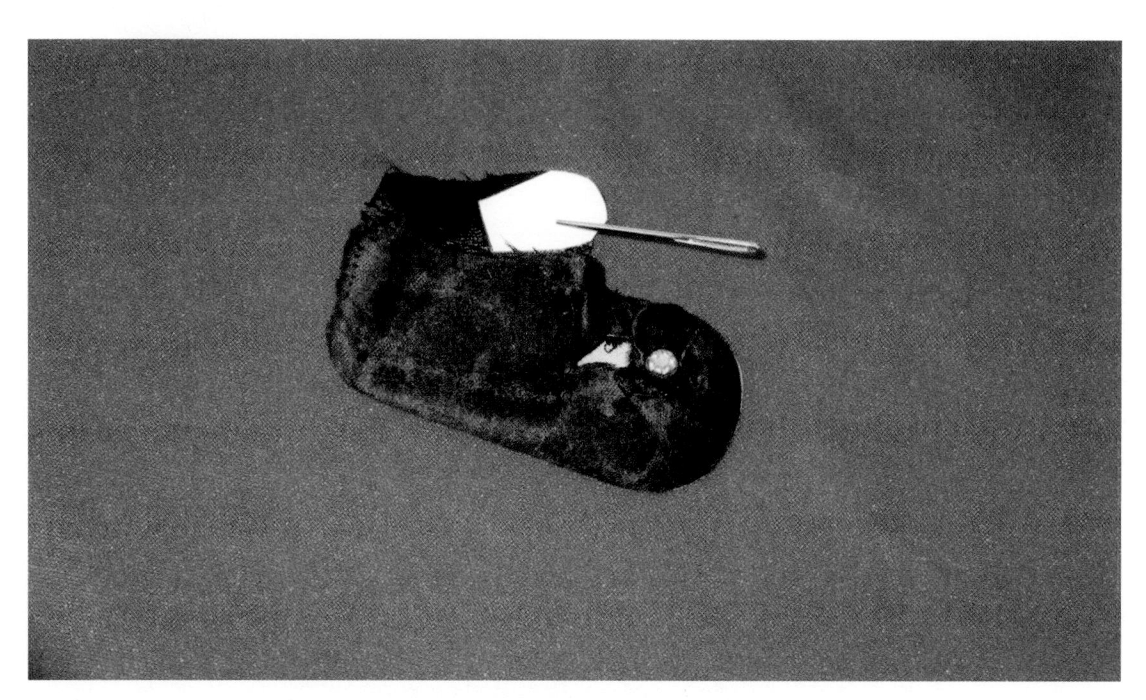

STEP #5. Make a new hole in the shoe strap.

8 If shoestrings are missing or frayed beyond use, silk ribbons can be used as elegant replacements. You can even fashion a matching hair bow from the same ribbon. (**SEE #8.**)

9 Doll shoes made of a material like old oilcloth clean up very nicely with doll cleaner.

10 Put vinyl doll shoes in a margarine tub (or a small bowl) with liquid bleach in the water and Palmolive® dishwashing soap.* Let the shoes soak for a day or two - it will not hurt them.

11 Rinse them thoroughly and wipe them off.

12 Using the same water and a toothbrush, scrub old doll shoes that are made of an oilcloth-type material. The water will not hurt them, except for the soles. If the soles are cardboard, then you need to be careful not to wet them as they will fall apart.

13 Many Madame Alexander shoes were made with layers of paper or cardboard stuck together and covered with satin material. If they have been wet, they may have separated and expanded between the layers.

14 Using a toothpick dipped in glue, work the glue in between the layers and press them together.

15 Hold them with a paper clip or a clothespin to dry before going onto the next section of the shoe!

*Palmolive® dishwashing soap is recommended to use with bleach because it blends properly with the pH factor and does not make fumes.

STEP #8. A new ribbon for the tie.

Restoring Leather Fringe and Chaps

The leather used for doll clothing tends to be very thin and is cut into very small pieces when fringed. Therefore it has quite a tendency to curl.

1. To steam leather fringe and chaps, **do not** use a steam iron. Use either a hand-held steamer or boiling water. (The boiling water method is recommended as it is easier and faster. With the hand-held steamer method, you hold the garment in one hand and the steamer in the other while you have both hands free to work on the garment with the boiling water method.)

2. Boil water on the stove until the steam is rolling well.

3. Hold the garment above it and let the steam penetrate it.

4. After steaming the leather fringe or chaps, smooth it with your hands.

5. Cover it with a tea towel or a piece of fabric. Do not use a terry cloth towel, as it will make pitted designs in the leather as it cools.

6. Put books on it to weigh it down.

7. Leave this for several hours to cool and dry. The fringe should come out flat and straight and the chaps will also.

Restoring a Felt Hat

In order to work on restoring a felt hat, you will need something on which to mount the hat so that it fits as snugly as if it were on the doll's head. You can use the doll itself if you cover her completely (head and all) with a plastic bag or plastic wrap. The whole procedure will take time and patience.

1. Firmly place the hat on whatever you are using to mount it on.

2. If the crown is droopy, stuff it really tight with polyester fiber filling or colorless tissue paper so you will have the correct shape.

3. Using a fabric finish (Magic Sizing is recommended), spray the hat — the crown area first — and let it dry.

4. When dry, put a sandwich bag over that part and spray the brim on both sides - top and bottom.

5. Remove the hat and press the brim with an iron.

6. If you want it to curl up a little, spray a little more sizing and curl it with your hands.

Restoring Fabric

Satin

1. **Always** press satin on the **wrong side**.
2. Usually the skirts are shaped well enough that you can get the iron inside of them to press.
3. Use a sleeve board if you have one.

Taffeta

Since I have not had any luck sending a taffeta garment to be dry-cleaned (they do not know how to press little garments), this method is recommended.

1. **Always** press taffeta on the **wrong side**.
2. If you have a little hand steamer, hang the garment up and steam it. Try to smooth it with your hands.
3. If you do not have a steamer, boil some water on the stove. Then hold the garment just above the pan to let the steam penetrate it.
4. Quickly lay it on a smooth towel to smooth it with your hands.

Antiquing Your Fabrics

You have looked and looked for some fabric to make a certain item of clothing for your doll and you want it to appear original. You have gone through your stash of material and you have searched garage sales and thrift stores for some garment that you could cut up and use, but all to no avail. There is one more alternative. Antique the fabrics that you have!

Tea dyeing has been very popular for quite some time, but you just want your material to look old – not dirty and worn.

Voile is a soft bridal fabric that looks and feels like the baby dresses you find on many of the old composition dolls. Plus it has enough cotton in it that it takes any color dye. Try this method on some other old fabrics such as old shirts and you will be surprised at the results. Play with fabric dyes like you would paint or watercolor. Brewed coffee and tea could also be used.
(SEE PHOTO BELOW.)

Follow the instructions on the fabric dye package and then try these colors:

- **Mauve:** use white voile; mix Rose Pink and Cocoa Brown.
- **Gray Blue:** use white voile; mix Baby Blue and Cocoa Brown.
- **Deep Ecru:** use white voile; mix Ecru and a touch of Cocoa Brown.
- **Pale Peach:** use white voile; mix Pale Pink and Ecru.

You note that Cocoa Brown is included in several of these color combinations. It is a wonderful color to add to most any combination to make it look old. Mix up several colors along these lines, including some of your own mixtures, and store them in glass jars. Then when you have a need to age something, just put a little of this and a little of that, add it to water and dip it! If you do not like the color you get, bleach will take it out and you can try again!

Dye for fabric
(to make it look old)

Cloth Dolls

4 Let the body soak for a few minutes.

5 Work on the worst areas first, usually the hands, because people tend to pick up a doll by the hands. Continue until the whole body is clean.

6 When the body is clean, rinse it well under running water and roll in a towel.

7 Press down hard to take out the excess moisture and hang to dry. **Do not** put it in a dryer as the fabric might shrink!

8 When it is dry, iron it completely and carefully. Spray starch will give it some body. Be sure to iron the fabric dry if you use starch.

9 Return the stuffing.

10 Carefully stitch all the openings back together by hand.

Composition Dolls

Composition was used first to make doll bodies for bisque head dolls. Some of them were a papier-mâché type of compressed material and some were a composite of wood shavings, sawdust and other materials. Those first bodies were often ball-jointed which allowed the dolls to have bent limbs so they could sit or pose their arms and legs. Sometimes a part of an arm or leg and the actual "ball" in the joint was solid wood, but painted to match the composite parts.

Composition was always painted a flesh tone and it varied in color from one company to the next. Also, the quality of the body itself as well as the paint used varied. Some composition dolls have lasted very well through the years while others have split at the seams and/or crazed, depending on the workmanship. Another factor is how well the doll was cared for through the years. A composition body or a doll completely made of composition that has been well cared for, probably has fared much better than one put away in an attic or garage where the temperature, humidity and other environmental factors were not controlled.

Heartbreak occurs when a doll, unpacked after thirty-five or forty years, is discovered to be all crazed and possibly cracked open in the corners of her eyes. She may be split open where there was once a seam. The cracks may be surface crazing in the paint or sometimes the paint may be lifting or peeling off altogether. The crazing may even go deeper than the paint down into the composition itself.

This is when the owner will seek someone to repair the doll. Even though, in many cases, the doll is unmarked and, therefore, possibly manufactured by a less well-known company and/or it may not be a particularly valuable or "collectible" doll. It is, nevertheless, valuable and important to its owner who, no doubt, has a strong sentimental attachment to it.

When the owner asks for an estimate to repair the doll, that estimate should be based on the time and work that it will take to do the restoration necessary. The value of the doll itself should not enter the picture at all. Restoration work should never reflect any difference in the quality of the doll to be restored. When these issues are adhered to, there is harmony between the owner and the restorer.

Restoration and Cleaning

To properly accomplish restoration of composi-tion dolls, you need a room set aside from the rest of your home or shop. This room should be very well lit and ventilated enough to allow painting to be an ongoing activity. You will find that this work is very messy involving lots of sanding, grit, dust, chips of paint, cans of paint and thinner, not to mention smelly brushes and other items.

I quickly discovered that my sewing area with the ironing board as a workbench was not the place for composition restoration. I covered and uncovered things for weeks so I could switch from one type of restoration to another. Before long, I decided to turn the composition restoration over to someone else and continue with the other part of doll restoration.

Therefore, the instructions given here are the very basics of composition restoration.

1. Chip off all the paint that is loose.

2. Fill the crevices. Auto body putty is recommended as a filler. It is easy to work with and easy to sand. A tablespoon of the putty and a drop of the cream hardener go a long way. When you purchase it at an auto parts store, ask the employee how to mix it. It needs to be used immediately, so have everything ready when you mix it.

3. Sand the doll smooth so you cannot tell where you have filled. This can take several days on just one leg or one head, as it must be done in layers and with lots of patience.

4. Prime it all. The primer has to dry thoroughly before you can sand it.

5. Sand lightly so no brush strokes are visible in the primer coat. You may find that what you thought was all filled and smooth will need more filling and sanding! If so, go back to the filling and sanding step each time.

6. When you feel that it is done to perfection and primed again, you are ready to paint.

7. Use a flesh-colored paint. You will find that the shades of flesh-colored paint differ a great deal from doll to doll. In addition, some have faded over the years. If you are trying to match the rest of the doll, take her to your paint store and have them mix paint to match.

8. Use very thin coats of the color — about six or more.

9. Let the paint dry a full twenty-four hours between each coat.

10. When all is done and you can see no brush strokes in the color (because the layers are so thin), seal it with a sealer. Of course, if you are working on the head and face, the features must all be painted on before you apply the sealer.

11. Let everything dry thoroughly before restringing the doll.

NOTE There are many steps to completing composition repair and whole books have been devoted to it. This includes learning to mold fingers where some have broken off (detailed instructions for repairing or replacing fingers are given in the section on "How to Make New Fingers" in this chapter), and how to sculpt the lines in a foot to make it look like the original one, with toes that have toenails and other features. If you want to learn more about this area of doll restoration, then you need to do more research on extensive and detailed composition restoration.

Crazed Paint That Is Not Openly Peeling Off

Crazing usually shows up first on the exposed parts of a doll. Her face, arms and legs are where she is handled and played with, therefore they are more prone to crazing. The parts of a doll that have been covered with layers of clothing are in much better condition than the exposed areas. For such a doll, I would recommend cleaning it with a doll cleaner.

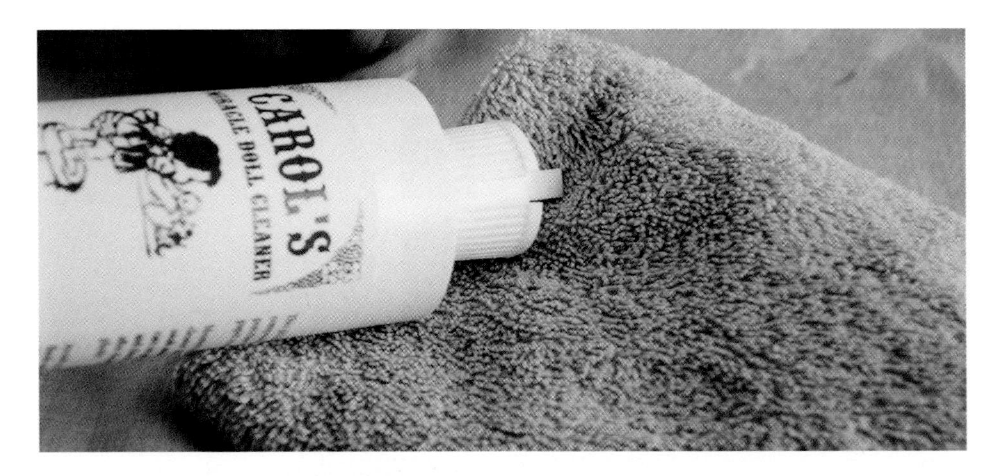

A. Applying cleaner to the washcloth.

B. Cleaning the doll.

How To Make New Fingers

This method will work on almost any kind of doll including composition, vinyl and rubber.

1 Push a straight pin very securely into the hand where you want to "make" a new finger. (**SEE #1A & #1B.**)

2 Use needle nose pliers to cut the head of the pin off 1/8in (.31cm) shorter than the finished finger should be.

3 Look at her other fingers on both hands and determine if the finger needs to be straight or curved.

4 If it should be curved, use pliers to bend the end to curve the pin.

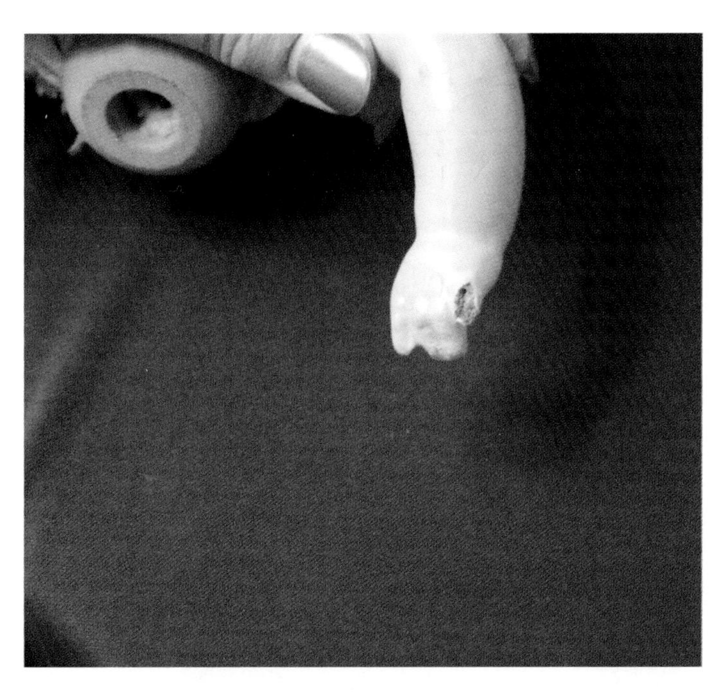

STEP #1A. A composition doll missing a little finger.

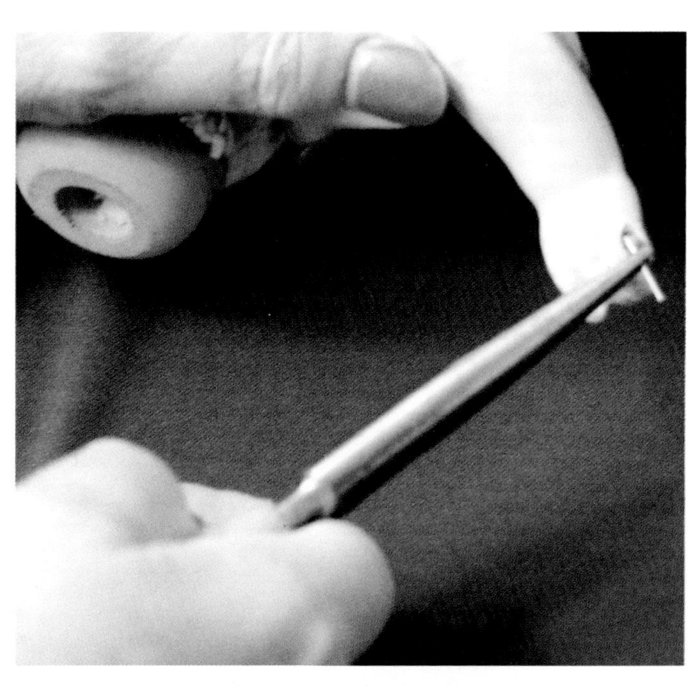

STEP #1B. Push a straight pin into the hand. Cut off the pin to desired length for the finger.

5 You will need the acrylic fingernail powder and liquid products that nail salons use for sculpted nails. These can be purchased at a beauty shop supply store in small sizes of each one. (SEE **#5.**) You will also need a small inexpensive brush. (If you are not familiar with how fingernails are sculpted, go to a nail salon and ask to be a spectator for a few minutes. You will need some practice.)

6 Pour some liquid out of the bottle into a very small container.

7 Cork the bottle as it will evaporate.

8 Dip just the tip of your brush in the liquid. Remember, a little goes a long way. (SEE **#8.**)

STEP #5. Nail sculpting products and tools.

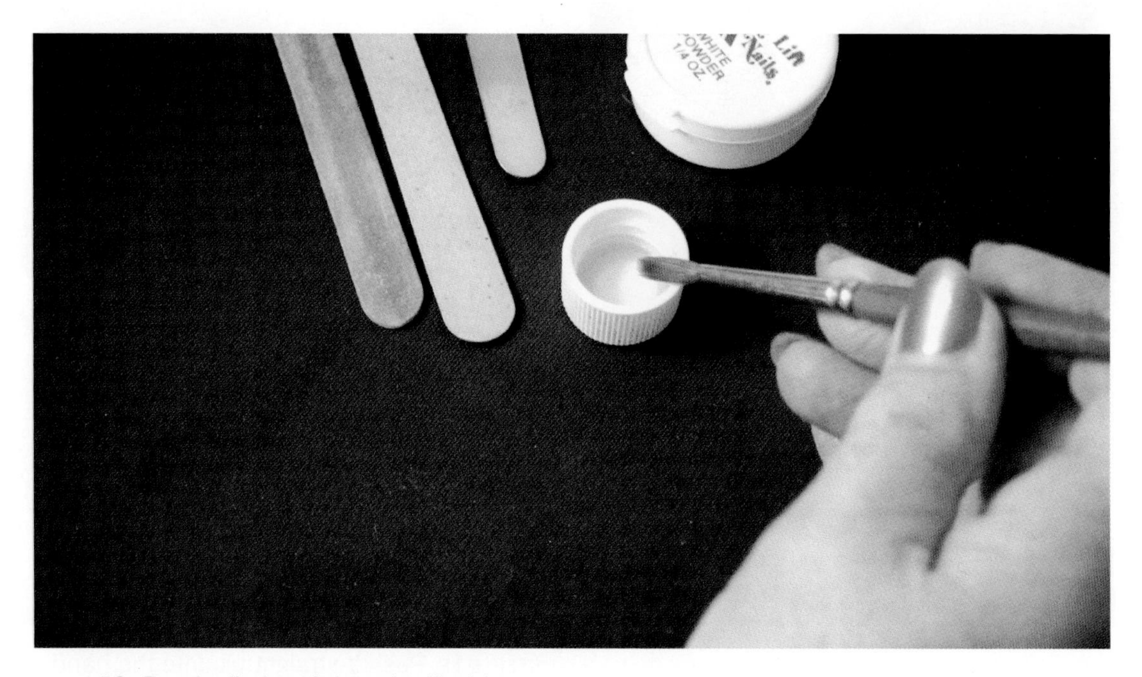

STEP #8. Barely dip brush into the liquid.

9 Then dip it in the powder. (SEE **#9.**)

10 Start building the finger on the pin.

11 Fill the crevice where the pin is, so you cannot see the inside of the composition, which looks like pressed wood. (SEE **#11.**)

12 If you have too much liquid on the brush you will use much more powder than needed in each step and it will likely be so runny that it will not stay where you put it. Practice is the only way to master this process.

13 Keep building on the pin out to the end until it is as large as you want the finger to be. You may have to let it dry in between layers in order to build it up as much as is necessary. Make it a little larger than you want the finished finger.

STEP #9. Barely touch the wet brush to the powder.

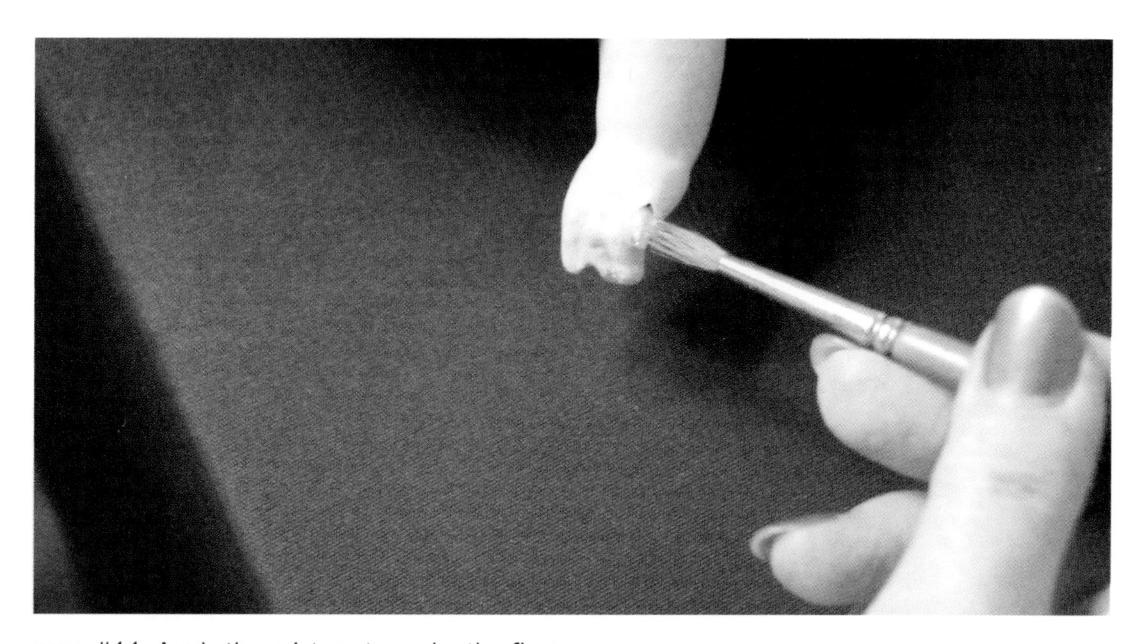

STEP #11. Apply the mixture to make the finger.

14 Let it dry in order to become solid.

15 File your new finger with an emery board much like you would use sandpaper. (SEE #5.) Sculpt it to resemble a finger or thumb!

16 You can use an X-acto knife, a single edged razor blade or even the point of your paring knife from the kitchen to make the creases on the knuckles.

17 File a place on the top side a little flatter than the rest of the finger and carve a little to make it appear as though she has a fingernail.

18 When you are happy with your work, brush off all the dust from the filing.

19 Coat your new finger with clear fingernail polish or perhaps a base coat.

20 When it dries, paint it with a coat of flesh-colored paint to match the rest of her body.

NOTE The acrylic nail sculpting material can be used for a number of repairs other than fingers. I have used it to repair a bisque head which was in pieces with some pieces missing. After it was all glued together, the gaps and cracks were filled in with this material, following the previous instructions. If this material is used to repair bisque or china (or even a figurine), be sure to explain to the customer that this will not be an invisible repair.

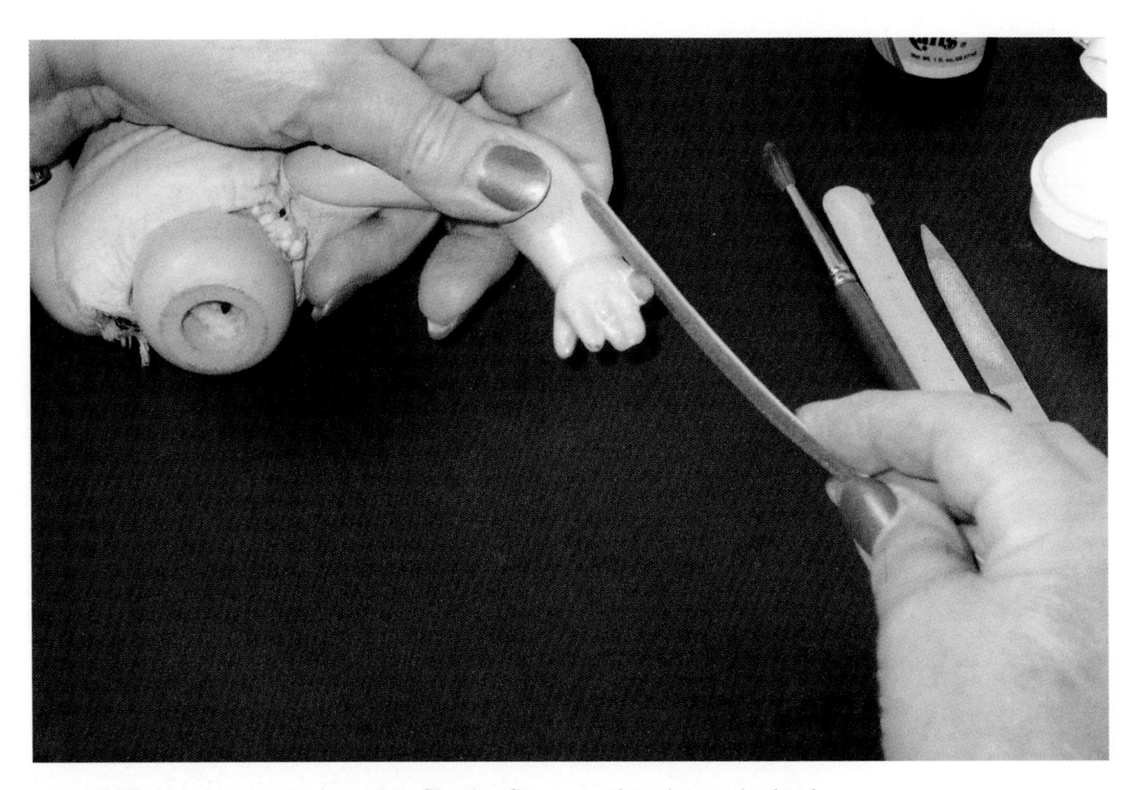

STEP #15. Use an emery board to file the finger to the shape desired.

Tips for Certain Kinds of Dolls

Caring For "Magic Skin" Dolls

In the 1940s and 1950s, doll companies tried methods of making dolls more "lifelike," one of which was to use "Magic Skin" for doll bodies. These "Magic Skin" dolls were a delight to the little mothers who received them because they looked and felt like they were real babies! Most of them had hard plastic heads, but their bodies and limbs were made of "Magic Skin." Others had the hard plastic heads and cloth bodies, with only the limbs of "Magic Skin."

When those dolls are taken from storage, the hard plastic heads are in wonderful condition! However, the "Magic Skin" parts of the dolls are in various states of deterioration. Some are found with completely black "Magic Skin." Some look like they have a case of measles with little round marks and holes all over the "Magic Skin" and the terrible flakes of green and black shredded foam stuffing show through. Many times the clothing is discolored as a result and often the "Magic Skin" and the stuffing have adhered to the clothing. The "Magic Skin" also seems to "balloon." That is - it swells during this reaction and the whole doll looks swollen. Sometimes the fingers are twice their normal size! This is quite a shock to many doll owners who thought they were storing their beloved childhood dolls in a safe place and in good condition.

To the best of my knowledge there is no cure for the "Magic Skin" problem. There seems to be a chemical reaction in the makeup of the "Magic Skin." I have made many new cloth bodies and attached either composition or vinyl limbs to the bodies in order to use the original hard plastic head that seems to be in perfect condition. This is why your "parts department" is so valuable. Many times you have a doll in your parts department that you can use for the body or limbs or both in order to restore another doll.

There is one thing I found that will help just a little if the deterioration of the "Magic Skin" has not gone too far. If you have just a slight split such as around the neck where the skin is covering a wooden neck button, get out your Band-Aids®! The Band-Aids® that are flesh-colored and look and feel thin like skin are the ones that you should use. Cut off the tape part of the Band-Aid® and use it to cover the split. It sticks very well to the "Magic Skin" and will hold together for as long as the "Magic Skin" lasts.

Caring for Cabbage Patch Kids

***Cabbage Patch Kids* were created to be played** with. It stands to reason that the cloth bodies will become very dirty and eventually wear out long before the vinyl parts. (SEE ILLUSTRATION ON THE RIGHT.)

A very dirty *Cabbage Patch Kid* can be put in the washing machine with a load of towels and will come out clean – clothes and all. It can also be put in the dryer with a load of towels.

If you need to replace the body on a *Cabbage Patch Kid,* simply find a replacement body, take the head off the body that needs to be replaced, and secure it to the replacement body with plastic cable ties.

Caring for Rubber Dolls

Wonderful rubber dolls were made both before and after World War II. Many of us had these dolls as children and tried to store them as keepsakes or are collecting them now. After many years, however, the rubber really shows signs of deteriorating. The best advice to preserve these dolls is to store them in a very even, moderate temperature and in an area where it is neither too humid nor too dry – not in the garage or the attic.

There is not any sure way of preventing the deterioration of the rubber but there is one method which can be helpful. Warm the limbs with a hair dryer, next to where they join the body. Pull one of the limbs out very carefully so as to not break the opening. Stuff each limb, the head and the body with polyester filling. This will help support the rubber and will hold it in the proper shape. A rubber doll with a collapsed body and/or limbs is usually beyond using this method. If you try to stuff the body and/or limbs, you will most likely cause a break or tear in the armhole or leg hole of the body when you try to remove it.

Saucy Walker and the Problems with Her Eyes and Head

The eyes in *Saucy Walker* dolls are notorious for coming loose. They are usually clamped onto a protrusion inside the head that is just above the nose.

1. Test first, holding the eyes inside the head to make sure they will fit correctly.

2. If you have needle nose pliers, you can reach in and clamp it tight again.

3. If you do not have pliers like that, you can lay her head face down. Put a lot of glue on the clamp after you have it in place.

4. Spread the glue all around on the inside of the head around the clamp.

5. A word of caution: Do not let the glue touch the eyes or their mechanism in any way. You want the eyes to work freely. Most *Saucy Walkers* were made with flirty eyes, which rolled back and forth across her head.

6. Lay her face down to dry. When these dolls were made, the eyes were clamped onto the inside of the head before the front and back of the head were adhered together.

7. Now you need to put her head back on. This is not an easy task and you really need two people to do it. There is a hook at the top of *Saucy Walker's* neck. (**SEE #7.**) That hook had a rubber band on which there was another hook to fit over the metal rod up in her head. This allowed her head to turn when she walked. You have a very small space in which to work in order to reattach her head.

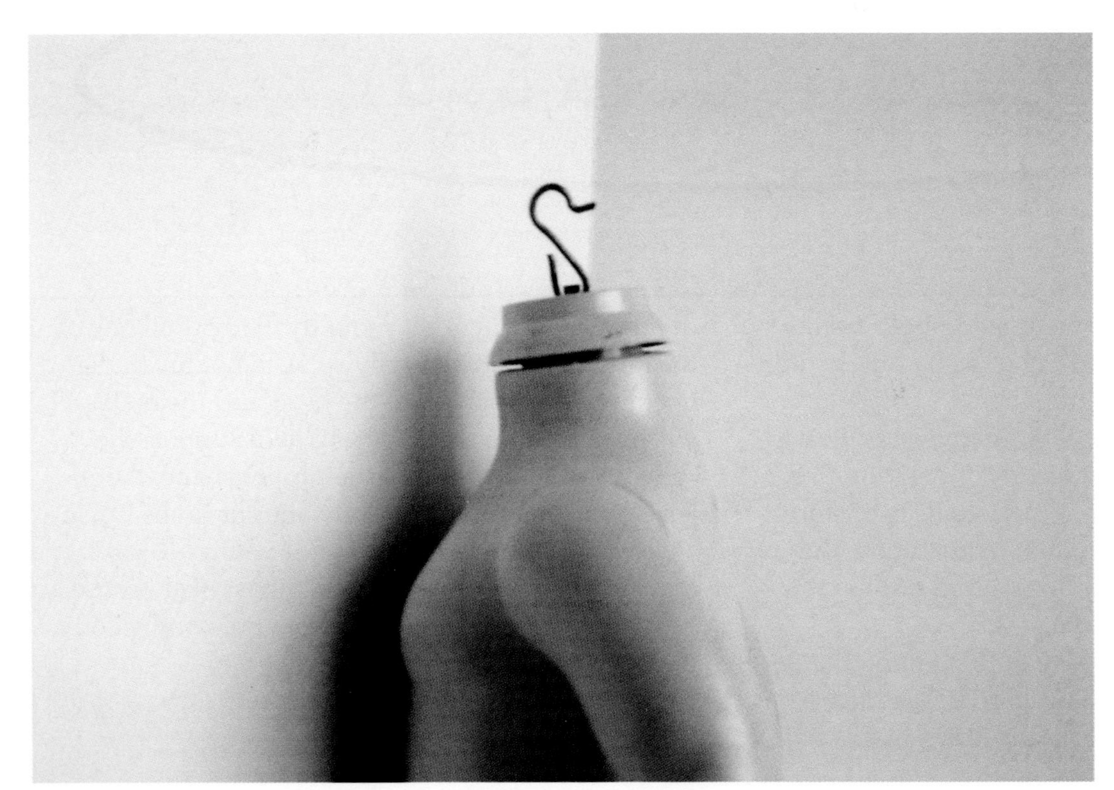

STEP #7.
Hook on the neck of *Saucy Walker* doll.

8 Cut a piece of restringing elastic about 3in (8cm) long.

9 Fold it in half and clamp the ends together very tightly. (SEE #9A.) This can be done with a hog ring and pliers. If you do not have those, you can use wire. (SEE #9B.)

10 Wrap it very tightly around the ends of the elastic to form a loop.

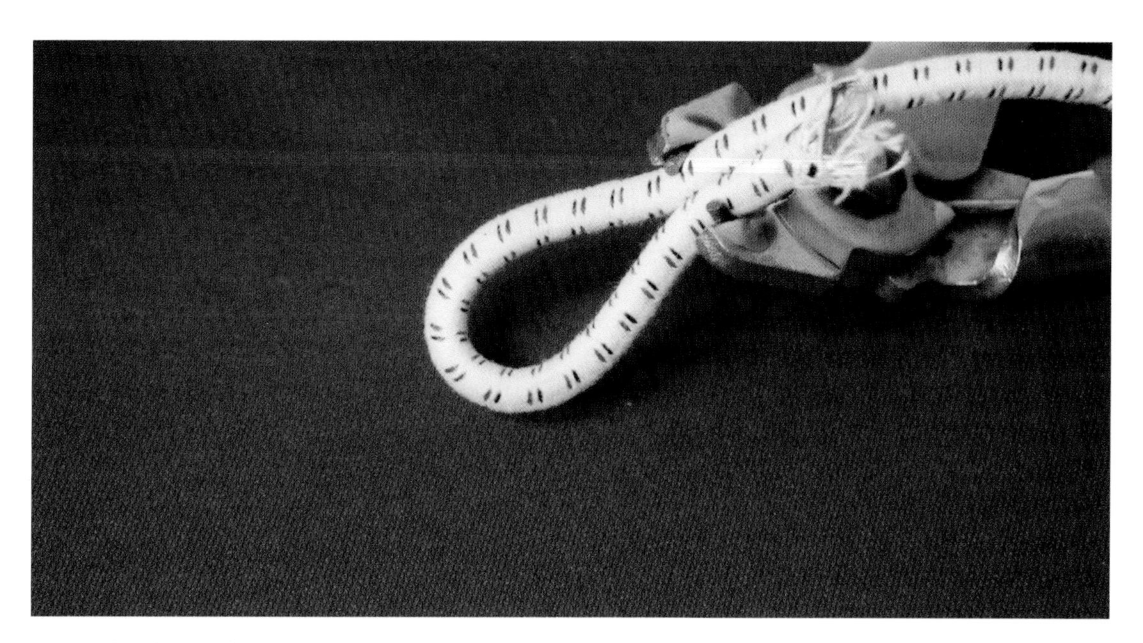

STEP #9A. Clamp the elastic loop with a hog ring.

STEP #9B. Or, use wire to twist on the elastic to make a loop.

11 Put a hook through this loop and then hook it up inside the head on the metal rod. (SEE #11.)

12 Now, with the help of another person, grab the elastic with a large crochet hook or buttonhook. (SEE #12.)

13 Pull it down and secure it on the hook on *Saucy Walker's* neck. (SEE #13.)

14 This needs to stretch very tight. You do not want her head to wobble, but to fit tightly against her neck so that her head will turn when she walks. You may have to do this several times before you have complete success. It is difficult to pull the elastic down. I have even resorted to opening the hook that is attached to *Saucy Walker's* neck so that it barely has a "hook" on it. Once through the loop, re-bend the hook so the elastic will stay on it.

15 When you have finished, if her head is not as tight as it should be, there are two things you can alter. Use a shorter, smaller hook up inside over the rod or use a shorter piece of elastic to double and clamp.

16 Then you have to start all over.

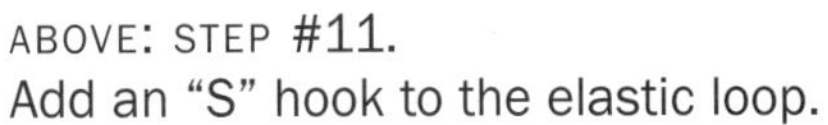

ABOVE: STEP #11.
Add an "S" hook to the elastic loop.

ABOVE RIGHT: STEP #12.
Pulling on loop with a crochet hook or other hook.

RIGHT: STEP #13.
Place the "S" hook on the neck through the elastic loop.

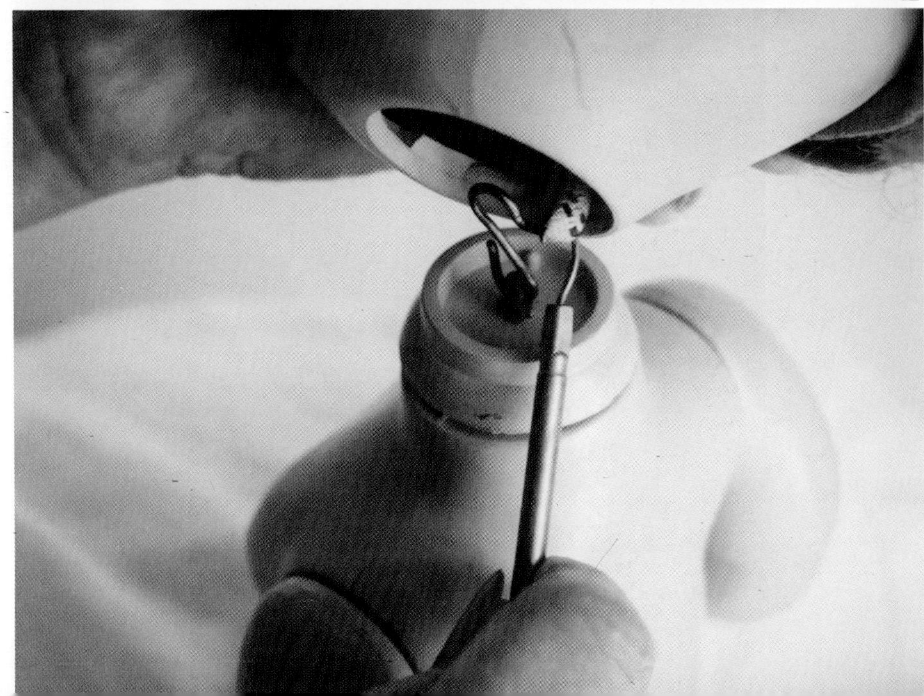

Shirley Temple Dolls

***Shirley Temple* dolls were first made in the early** 1930s of composition. Since then they have been made of hard plastic, vinyl and porcelain. Designed by Bernard Lipfert, the "father" of many of the more popular dolls from that era, the *Shirley Temple* dolls were extremely popular and the most widely copied doll of its time. The only authorized *Shirley Temple* doll was made by the Ideal Toy Corporation. While the most popular doll was the 13in (33cm) size, it was available in a number of different sizes.

Because of their popularity and their age, many of the composition versions are now in need of restoration work. Paint finish on the composition *Shirley Temple* dolls included a matte to a rosy gloss finish. Some have considerable eye shadow that is gray to black. Some dolls have a single eyelash line above as well as below the eyes along with a few eyelash strokes. Teeth ran the gamut from very well molded to just fair. Eyes could be a pale golden honey to a true brown, or a metallic hazel to a true green. Wigs were made of mohair and were pale golden yellow, ash blonde and strawberry blonde.

To clean and restore composition, vinyl and/or porcelain *Shirley Temple* dolls, as well as the clothing, please refer to the appropriate chapters for detailed instructions.

Vinyl Dolls: Repairing a Split

Vinyl can be repaired with Shoe Goo. (SEE PHOTO ON THE RIGHT.) It is the silicone glue that is used most on tennis shoes and it can be found in many department stores. Put it on and then somehow tape or clamp the split together. Leave it taped or clamped for days until it completely adheres to the split. Such a repair will not hold up under play conditions.

Glues

Use of Glue

When your doll has popped off a finger (and you still have it), most can be reattached. For doll fingers that are rubber, use a silicone cement or glue. (Shoe Goo is recommended.)

If the finger is hard plastic, then you can use super glue. The gel type super glue (sold in tubes on a card) is preferred. The gel kind can be squeezed out on a piece of paper and used with a toothpick to apply it where you want it.

If a finger is completely off, give it a little help by putting something such as a straight pin in the finger and the other end in the hand where it fits together. This will give it a little support to hold it. Glue it as well.

Restoration of a Miss Revlon Doll

Here is the *Miss Revlon* who has been restored to her original look. She was shown in the "Getting Started in Business" chapter. She is surrounded by her clothing. She needed to be cleaned and shampooed and her clothing needed to be repaired and restored.

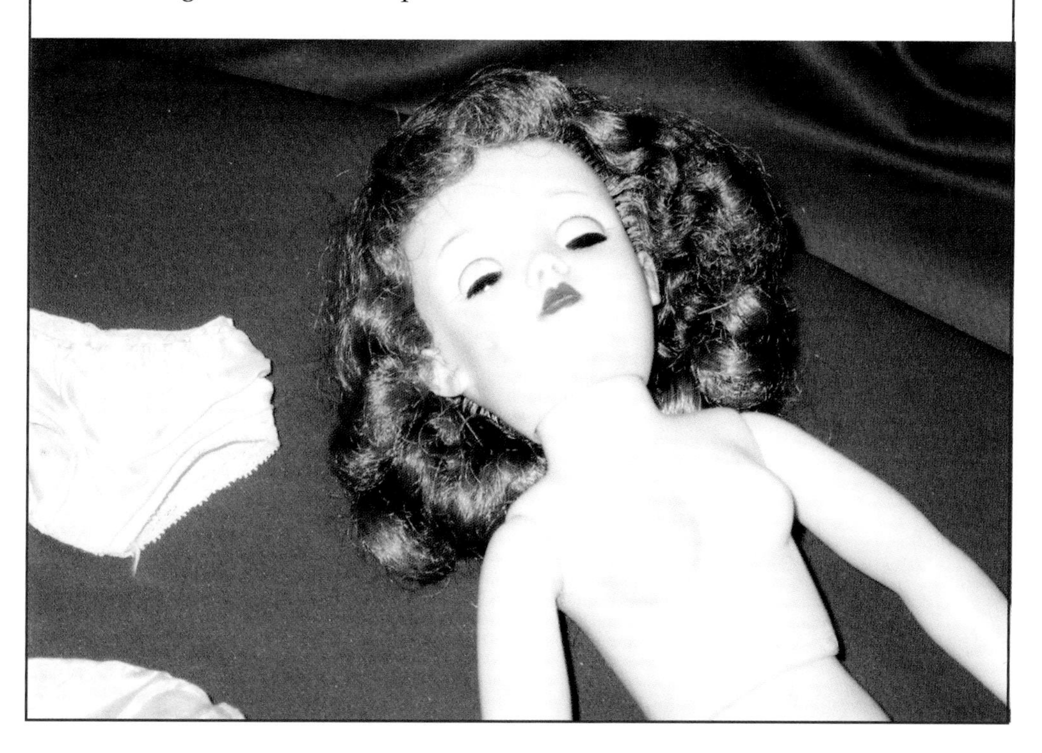

1 Her body was cleaned and her hair was shampooed with liquid fabric softener.

2 The restoration of her clothing began with her half-slip and panties.

3 The old elastic was removed from her panties.

4 It was also removed from her half-slip.

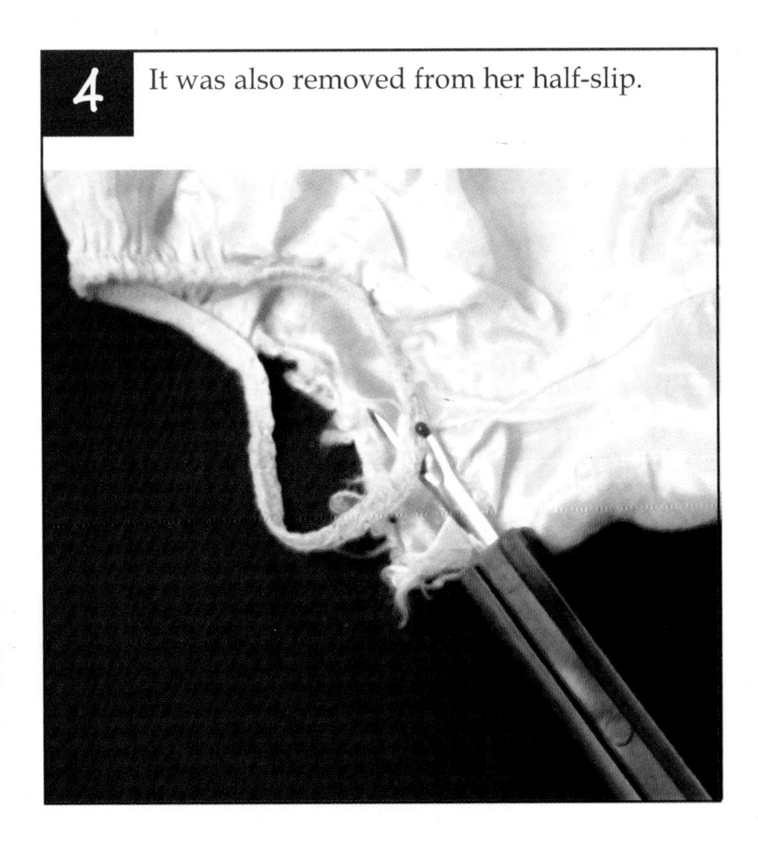

5 Pressing each item with a steam iron made a noticeable improvement.

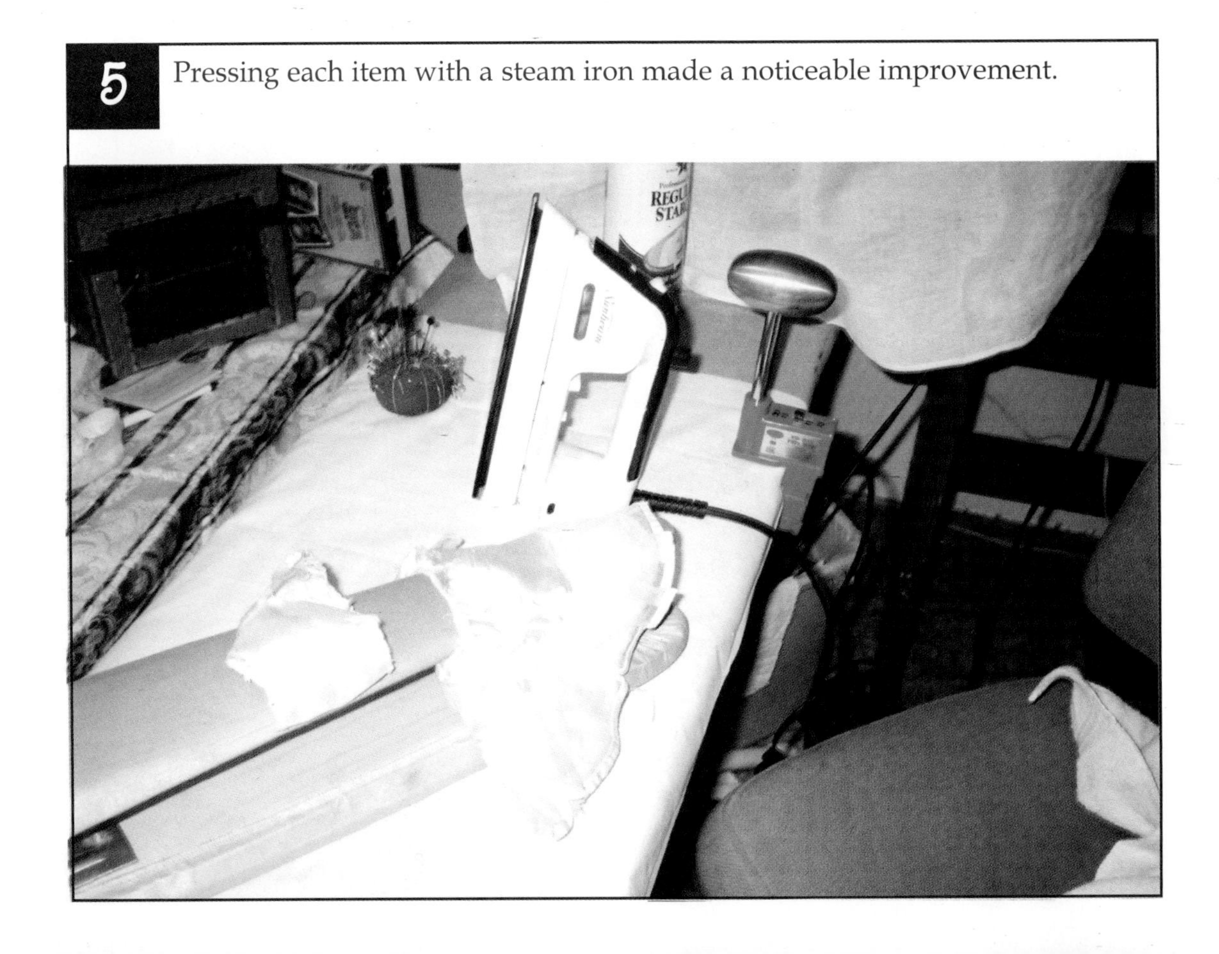

6 After cleaning her underwear, the elastic was replaced with new.

What an improvement!

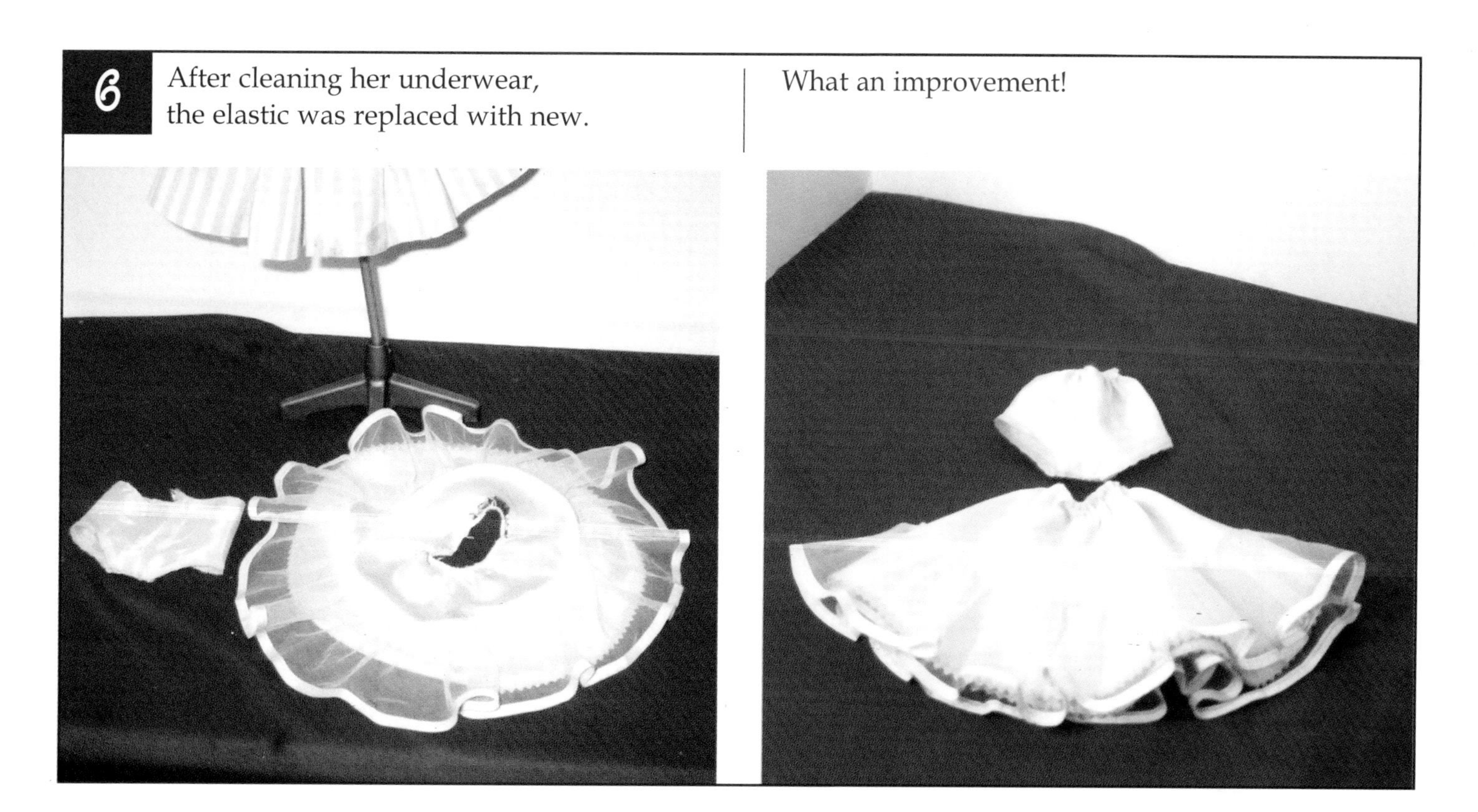

7 Next was her dress. All the loose threads and raveling were clipped from it.

8 The stain on the front hem of her dress was a bit of a concern. It needed special attention.

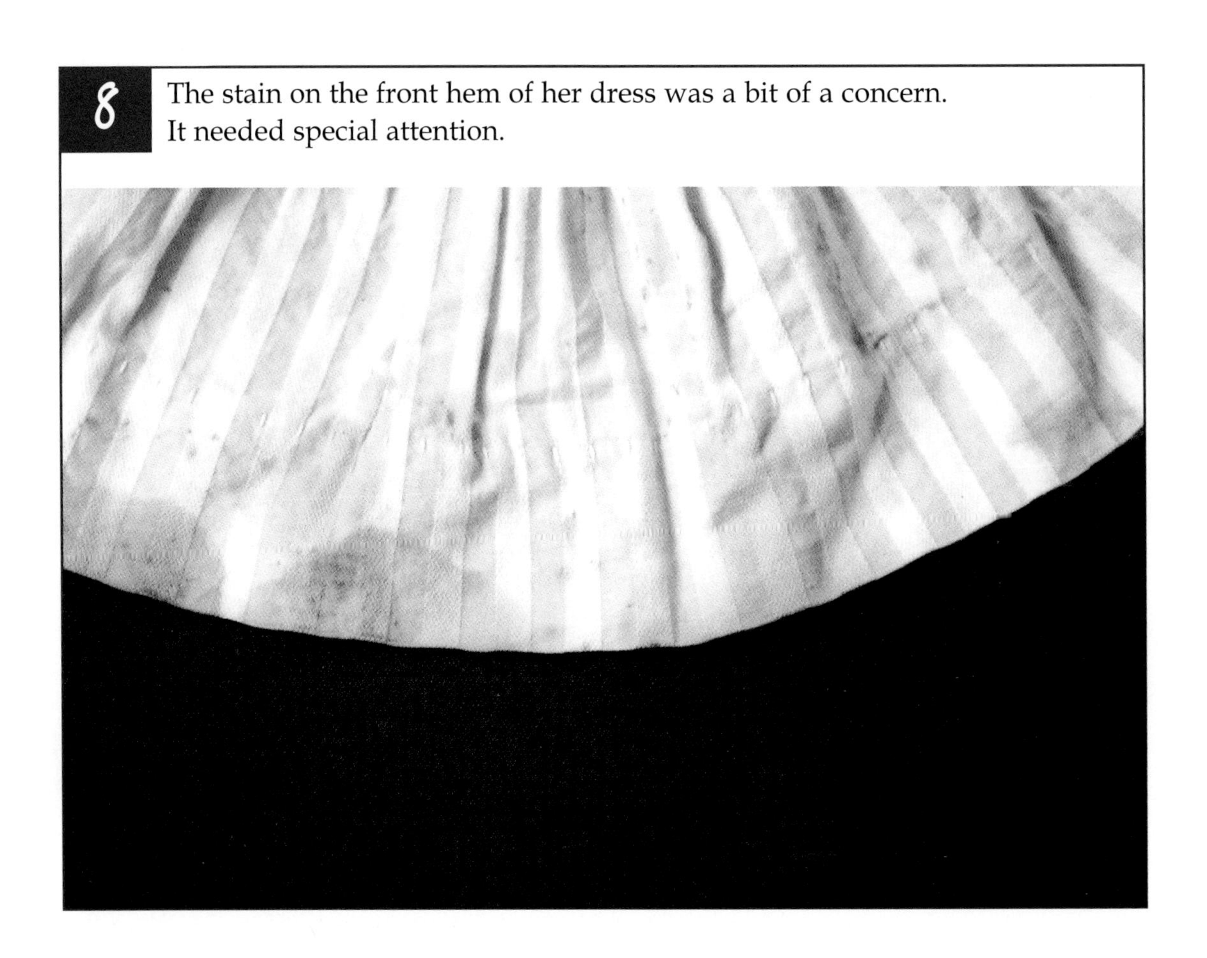

9 The stain was not completely removed. However, the rest of the dress is now nice and clean after washing in Palmolive® dishwashing soap and a little bit of bleach. The colors are even brighter.

10 Next comes the ironing. Positioning the bodice of the dress on the sleeve board makes a good work area. The entire dress was sprayed with spray starch and then ironed.

11 First, the bodice was ironed and then the skirt.

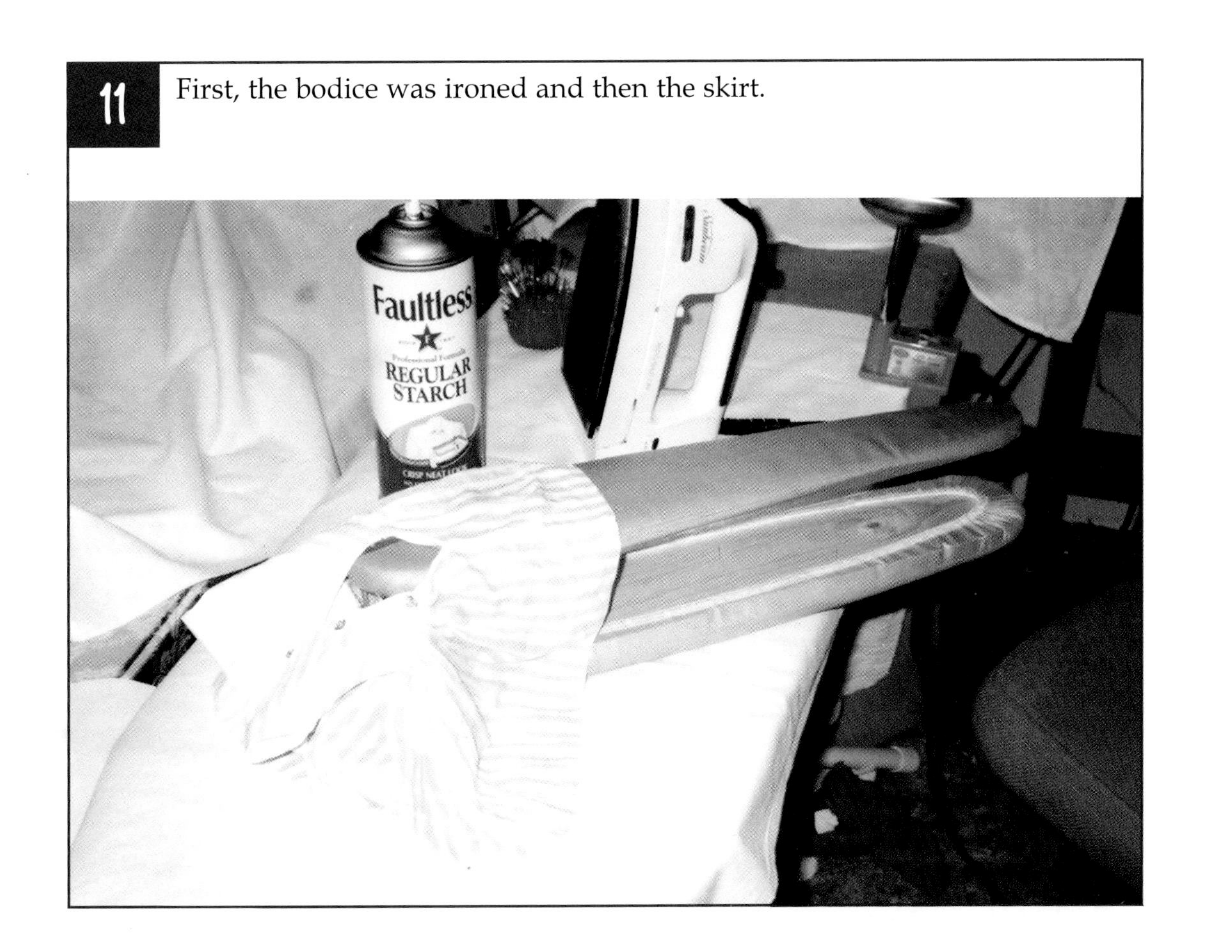

12 The dress was ironed slowly and exactly so the material could dry under the iron, as only then was it smooth and free from wrinkles. Ironing up into the gathers at the waistline makes the dress nice and full when worn.

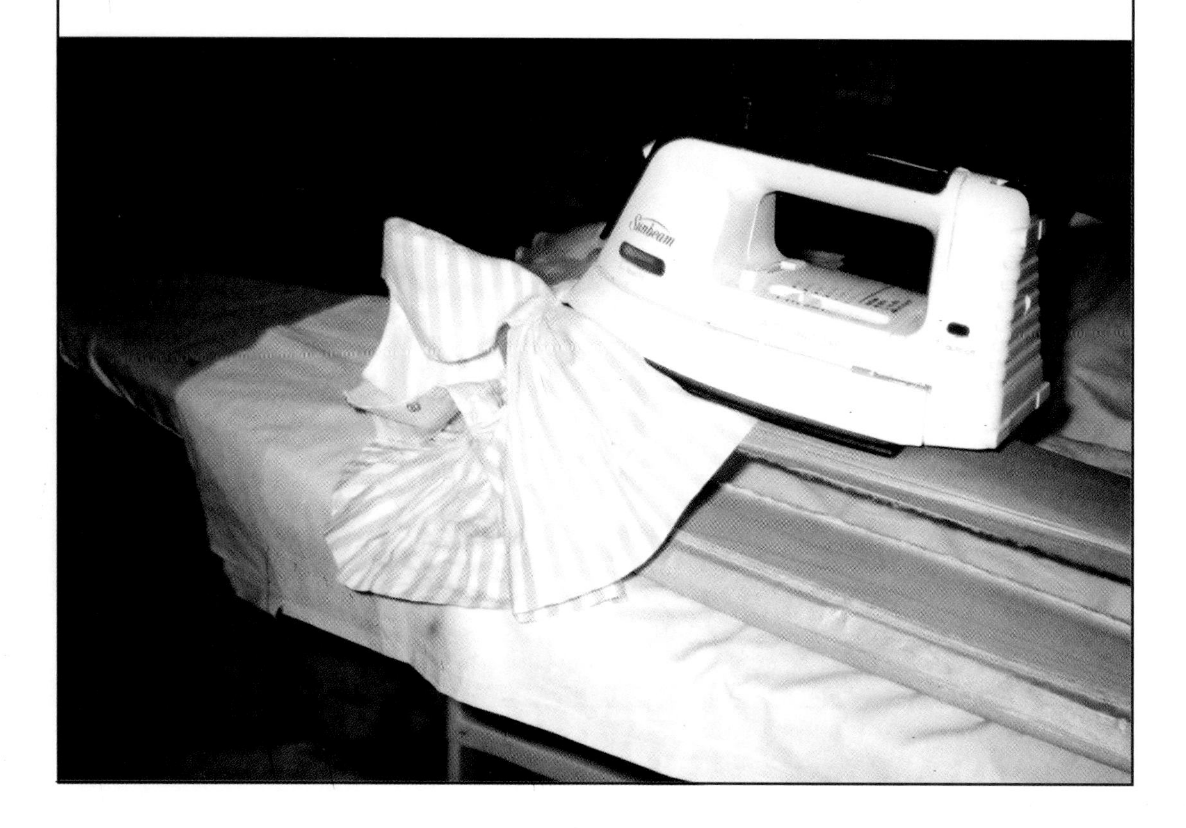

13 The waistline snap in the back was replaced. That completed the restoration of the dress.

14 After her hair was dry, it was combed into the original style. Her hair did not require any setting or rollers. It was simply allowed to dry.

15 With all of her clothing ready to wear, her hair combed and styled, she was ready to dress.

16 Her panties and half-slip fit her waist perfectly! Her dress was next.

17 Finally completed, she is wearing her newly restored dress. The rust looking stain was very stubborn. Though it was diminished to some extent, it could not be removed completely.

Her pearl-drop earrings are the only things missing from her original outfit. Replacement earrings can be fashioned out of some old pearls.

An illustration of two *Miss Revlon* dolls wearing identical dresses except that one is pink and the other is lavender.

*Restoring dolls is a most enjoyable
and fulfilling experience for me —
and I hope it is for you too!*

About the Author

Carol Lindberg, mother of three, grandmother of six and great-grandmother of one, has had a love affair with babies and children all her life and this has carried over into her love for dolls. As a matter of fact, her most beloved doll from childhood was *"Sweetie Pie,"* a cloth doll with a mask face, which she named after a composition doll by the Effanbee Doll Company which was popular about the same time she received her doll. The cloth body on the doll, which Carol still has, wore out many times and was patched and re-covered over the years. A few years ago, Carol was able to find a replacement for her *"Sweetie Pie,"* in much better condition and with its original body.

While her children were growing up, Carol's interest in dolls had to take a back seat, but once they were grown and left home, she was able to indulge once again her interest in dolls, developing a real passion for doll restoration and repair.

Acquiring numerous dolls in various stages of disrepair, Carol began to work on them. Utilizing the methods of restoration and repair she had learned, Carol honed her skills, perfecting her restoration and repair methods and techniques to the point that she even developed her own doll cleaner.

This book represents the culmination of many years of experience in doll restoration and repair methods which the author presents here for the readers' consideration.

LEFT.
The author, in 1940, and her childhood doll, "*Sweetie Pie*," a cloth doll with a mask face.

RIGHT.
The author's recently purchased replacement "*Sweetie Pie*" with its original cloth body.